Coastal Systems

Second edition

G000075041

The coast represents the crossroads between the oceans, land and atmosphere, and all three contribute to the physical and ecological evolution of coastlines. Coasts are dynamic systems, with identifiable inputs and outputs of energy and material. Changes to input force coasts to respond, often in dramatic ways as attested by the impacts of the Indian Ocean tsunami in 2004, the landfall of Hurricane Katrina along the Gulf Coast of the USA in 2005, and the steady rise of global-warming-driven sea level. More than half the world's human population lives at the coast, and here people often come into conflict with natural coastal processes. Research continues to unravel the relationship between coastal processes and society, so that we may better appreciate, understand, manage and live safely within this unique global environment.

Coastal Systems offers a concise introduction to the processes, landforms, ecosystems and management of this important global environment. New to the second edition is a greater emphasis on the role of high-energy events, such as storms and tsunamis, which have manifested themselves with catastrophic effects in recent years. There is also a new concluding chapter, and updated guides to the ever-growing coastal literature. Each chapter is illustrated and furnished with topical case studies from around the world. Introductory chapters establish the importance of coasts, and explain how they are studied within a systems framework. Subsequent chapters explore the role of waves, tides, rivers and sea-level change in coastal evolution.

Students will benefit from summary points, themed boxes, engaging discussion questions and new graded annotated guides to further reading at the end of each chapter. Additionally, a comprehensive glossary of technical terms and an extensive bibliography are provided. The book is highly illustrated with diagrams and original plates. The comprehensive balance of illustrations and academic thought provides a well-balanced view between the role of coastal catastrophes and gradual processes, also examining the impact humans and society have and continue to have on the coastal environment.

Simon K. Haslett is a Professor of Physical Geography and recently appointed Director of the Centre for Excellence in Learning and Teaching (CELT) at the University of Wales, Newport. His main area of research is coastal evolution and oceanography, including coastal upwelling, sea-level change, and the impact of high-energy events, such as tsunami and storms. He is dedicated to the Public Understanding of Science and has written extensively in newspapers and magazines, gives public lectures, and appears on television. He was very heavily involved in making a television programme on the Bristol Channel in 2004 for BBC Timewatch, entitled *Killer Wave of 1607*, which was screened in April 2005.

Routledge Introductions to Environment Series

Published and Forthcoming Titles

Titles under Series Editors:
Rita Gardner and A. M. Mannion

Environmental Science texts

Atmospheric Processes and Systems
Natural Environmental Change
Environmental Biology
Using Statistics to Understand The
    Environment
Environmental Physics
Environmental Chemistry
Biodiversity and Conservation, 2nd Edition
Ecosystems, 2nd Edition
Coastal Systems 2nd Edition

Titles under Series Editor:
David Pepper

Environment and Society texts

Environment and Philosophy
Energy, Society and Environment, 2nd
    edition
Gender and Environment
Environment and Business
Environment and Law
Environment & Society
Environmental Policy
Representing the Environment
Sustainable Development
Environment and Social Theory 2nd edition
Environmental Values
Environment and Politics 3rd Edition
Environment and Tourism 2nd Edition

Forthcoming
Environment and the City
Environment, Media and Communication
Environment and Food

Routledge Introductions to Environment Series

Coastal Systems

Second edition

Simon K. Haslett

Routledge
Taylor & Francis Group

LONDON AND NEW YORK

First published 2009 by Routledge
2 Park Square, Milton Park, Abingdon, Oxon, OX14 4RN

Simultaneously published in the USA and Canada
by Routledge
270 Madison Avenue, New York, NY 10016

Routledge is an imprint of the Taylor & Francis Group, an informa business

Typeset in Times New Roman and Franklin Gothic
by Keystroke, 28 High Street, Tettenhall, Wolverhampton
Printed and bound in Great Britain by
TJ International Ltd, Padstow, Cornwall

British Library Cataloguing in Publication Data
A catalogue record for this book is available from the British Library

Library of Congress Cataloging in Publication Data
Haslett, Simon K.
    Coastal systems / Simon K. Haslett. — 2nd ed.
        p. cm. — (Routledge introductions to environment)
    I. Title.
    GB451.2.H38 2008
    551.45′7—dc22   2008003336

ISBN10: 0–415–44061–0 (hbk)
ISBN10: 0–415–44060–2 (pbk)
ISBN10: 0–203–89320–4 (ebk)

ISBN13: 978–0–415–44061–5 (hbk)
ISBN13: 978–0–415–44060–8 (pbk)
ISBN13: 978–0–203–89320–3 (ebk)

For Sam, Maya, Elinor and Rhiannon

Contents

Plates

Figures

Tables

Boxes

Author's preface to the first edition

In 1998 I undertook fieldwork along the coast of northern Queensland in Australia, and at each beach I visited there were signs alerting visitors to various hazards. These included dangerous currents, stinging jellyfish, saltwater crocodiles, sunburn, and even falling coconuts! Being forewarned, I was able to enjoy investigating the various locations. However, there is a very serious message embedded in this: that coasts command respect. In recent years there have been a number of major coastal disasters resulting in thousands of casualties, including the Papua New Guinea tsunami in 1998, the Orissa storm surge in the Bay of Bengal in 1999, and the devastating flooding of the coastal lowlands of Mozambique in 2000 (some of these are discussed further in this book).

Around three billion people live in the coastal zone, that is half of the world's entire human population. It is no wonder then that coasts are under pressure. Indeed, it is a major challenge for the twenty-first century for humans to live and work at, and exploit, coasts in a way that does not damage the environment or deplete resources, and that we will pass on to our children that which we inherited from previous generations. If we are to achieve this goal of sustainable development, then we should be striving to understand better how physical and ecological coastal systems operate, and then work alongside these, and not battle against them.

It has been my intention in writing this book to introduce to the reader our current understanding of coastal systems, including the physical, ecological and human interactions, so as to raise an awareness of the diversity and sensitivity of these precious environments. To conform with the 'Introductions to Environment' series, this book has been written for first- and second-year undergraduate audiences studying coastal systems as a one-semester module. Out of necessity, therefore, the text proceeds from first principles, but I hope that there is sufficient detail herein to satisfy more advanced readers. To this end, I have carefully selected references for further reading.

I have been supported for many years in my oceanographic research by a large number of people and to all of them I am very grateful. In particular, I am indebted to my wife Sam and my children Maya, Elinor and Rhiannon, for their patience, encouragement and companionship in my oceanic endeavours, and to my parents for fostering my environmental interests from a very early age. My professional development has benefited greatly from my mentors Brian Funnell, John Murray, Rick Curr, Ted Bryant, Annika Sanfilippo and Allan Williams. Also, all my colleagues at Bath Spa University College have offered very sympathetic support to me whilst I have been writing this book, especially Paul Davies and Fiona Strawbridge who have been brave enough to embark on coastal research collaborations with me. I am also very grateful to Kevin Kennington, Andy Cundy and Chris Spencer for taking the time to digest the first draft of the text and to offer very constructive comments. Finally, thanks to Ann Michael and Casey Mein at Routledge for making the writing of this book as painless as possible.

Simon Haslett
Newton Park, Bath
April 2000

Author's preface to the second edition

In the opening paragraph of my preface to the first edition of *Coastal Systems* I stressed that coasts command respect and went on to mention some of the unfortunate coastal disasters that had occurred just before 2000, such as the 1998 Papua New Guinea tsunami and the 1999 Indian storm surge, both of which claimed lives running into several thousands. Events since 2000 show that such disasters are not isolated occurrences but contribute to an ongoing and terrible catalogue of coastal catastrophes. For then, in 2004 a tsunami affected the entire coastline of the Indian Ocean killing over a quarter of a million people, and in the following year the landfall of Hurricane Katrina in the United States caused death and destruction to areas around the Gulf of Mexico, showing that coastal disasters are not restricted to the developing world. Moreover, just as I was finishing the draft of this second edition yet another storm surge struck Bangladesh, again claiming thousands of lives – at the last minute I rewrote a section of the book to include this timely reminder of the respect that society should always pay to the coast, no matter where you are in the world.

This message underpins the societal value of studying coastal systems, allowing us to better understand how this complex environment developed through time, how it operates on different time and spatial scales, and what it may do in the future. Part of the future development of coastlines depends on human activity and the impacts of society, either through direct influence, such as quarrying coastal sediments, or indirectly through agents such as anthropogenic climate change. Indeed, this second edition has been timed to incorporate the latest predictions of the Intergovernmental Panel on Climate Change published in September 2007, which reiterate that not only may sea level rise as a result of global warming but that increases in storms and their severity are likely to exacerbate coastal erosion and flooding in many areas of the globe. This is an issue in which every individual on earth has a vested interest and is able to affect the outcome, for better or worse, depending on whether they are able to change their lifestyles to reduce greenhouse gas emissions or not.

I feel it is important for a textbook like this to arrest the attention of readers and to draw them into this highly interesting and often exciting field of study. Armed with the basic principles provided by this book I hope the student will be inspired enough to read further, undertake a field project, contemplate further study, and even consider a related career. The majority of reviews for the first edition, for which I'm grateful, have encouraged me to think that the format, scope and content of the book appeal to a wide audience of undergraduate students following a number of different discipline pathways, and for that reason I have not made any extensive changes to the book. My main aim for this second edition is to make it more appealing to the next generation of coastal science students by updating the ever-growing literature base and including examples that have occurred relatively recently and so would be familiar to students just entering university. I have also added further guidance to reading, which is becoming more and more bewildering with the expansion of online literature databases.

Since 2000 my coastal research has diversified somewhat and has benefited greatly through collaboration with other coastal scientists. I particularly want to acknowledge the pleasure I have had, and knowledge I have gained, at both ends of the analytical spectrum, through close collaboration with Professor John R. L. Allen (University of Reading) and Dr Ted Bryant (University of Wollongong). Whereas with John, we have been working on highly detailed analyses of estuarine sediments and landforms, sometimes at the sub-millimetre scale, to painstakingly piece together coastal evolution in relation to sea-level change over millennia, Ted and I have undertaken relatively broad-brush field studies of the impact of tsunami and storms along extensive coastal landscapes at single points in time. Such divergence of approaches and spatio-temporal contexts is very insightful and has shown me that one may never stop learning about the coast, even if you've written a textbook on the subject! I would also like to acknowledge other coastal colleagues to whom I want to signal my continued appreciation: Dr Andy Cundy (University of Brighton), Professor Roland Gehrels (University of Plymouth), Professor Jon Nott (James Cook University), Dr Chris Spencer (University of the West of England), and Dr Colin Woodroffe (University of Wollongong). My thanks also go to Andrew Mould and his team at Routledge for encouraging me to propose this second edition and for working with me on it, and to my colleagues of Bath Spa University for easing the task of writing and to the generations of students I have taught for helping to shape the content of the book.

Finally, I would like to endlessly thank my wife Sam and daughters Maya, Elinor and Rhiannon, for allowing me to indulge my interests and to follow this career, which often demands of me more than I'm paid to do and so takes time away that I could be spending with them. Thank you.

Simon Haslett
Newton Park, Bath
January 2008

Note on the text

Bold is used in the text to denote words defined in the Glossary.

Acknowledgements

The author and publisher would like to thank the following for granting permission to reproduce images in this work.

R. A. McBride *et al*. for Figure 2, reprinted from *Marine Geology*, Volume 126, McBride *et al*, Geomorphic response-type model for barrier coastlines: a regional perspective', pp. 143–159, 1995, with permission for Elsevier Science.

G. Masselink and C. B. Pattiaratchi for Figure 1, reprinted from *Marine Geology*, Volume 146, Masselink *et al*., 'Morphological evolution of beach cusps and associated swash circulation patterns', pp. 93–113, 1998, with permission from Elsevier Science.

J. R. L. Allen for Figures 2 and 5, reprinted from *Proceedings of the Geologists' Association*, Volume 104, J. R. L. Allen, 'Muddy alluvial coasts of Britain', pp. 241–262, 1993 and Figures 1 and 2, reprinted from *Proceedings of the Geologists' Association*, Volume 107, J. R. L. Allen, 'Shoreline movement and vertical textural patterns . . .', pp. 15–23, 1996, both published with permission of the Geologists' Association.

H. G. Reading, J. D. Collinson, and Blackwell Science for Figure 6.2 reprinted for the chapter 'Clastic Coasts' in the book *Sedimentary Environments*, 3rd edition, edited by H. G. Reading, 1996.

R. W. Young *et al*. for Figure 1, reprinted from *Australian Geographer*, Volume 27, R. W. Young *et al*, 'Fluvial deposition on the Shoalhaven Deltaic Plain, southern New South Wales', pp. 215–234, 1996, with permission form Taylor and Francis.

Natural England for permission to extract data from their Heritage Coasts Factfile web site at http://www.countryside.gov.uk/what/hcoast/heri_tbl.htm

D Briggs *et al*. and Routledge for Figure 1 (from the Box on p. 308), 1.6, 1.10, 3.5, 4.7, 4.11, 4.12, 4.13, 12.12, 16.1, 16.6, 17.1, 17.2b, 17.3, 17.4, 17.5, 17.12, 17.13, 17.14, 17.16, 17.17, 17.18 and 17.20 reprinted from *Fundamentals of the Physical Environment*, 2nd edition, 1997.

P. French and Routledge for Figures 2.3, 2.8, 2.10 and 2.11, reprinted from *Coastal and Estuarine Management*, 1997.

R. Kay, J. Alder and E & FN Spon for figures from Boxes 4.4 and 4.16, reprinted for *Coastal Planning and Management*, 1999.

C. Park and Routledge for Figures 11.25, 13.8, 15.4, 15.5, 15.12, 15.13, 15.14 and Table 11.9, reprinted from *The Environment*, 1997.

K. Pickering, L. Owen and Routledge for Figure 8.7, reprinted from *An Introduction to Global Environment Issues*, 2nd edition, 1997.

Introduction

The coast is a very special environment in that it is where the land, sea and atmosphere meet. Each of these contribute to the workings of the coast, making coasts very interesting, and yet a challenging subject to study. The coast is also the location of major human settlements, and human activity can have significant impacts on the operation of coasts to the point of environmental and socio-economic degradation. The study of coasts, therefore, is highly interdisciplinary, incorporating geology, physical and human geography, oceanography, climatology, sociology, economics, engineering, planning, management, and so on. It is perhaps one of the best examples of interdisciplinary environmental science.

It is clear that coasts are not isolated environments: they receive energy and material, process these inputs, and subsequently may lose them as output. Variations in inputs cause changes in the physical environment, for example an increase in wave-energy may enhance coastal erosion, or a decrease in nutrients may limit biological productivity, so demonstrating that all natural coastal operations are interlinked in some way. This describes a system, and the study of coasts is well suited to a systems approach.

The underlying theme throughout this book is the operation of coastal systems, explicitly demonstrated at times, and implied at others. However, in the real world the workings of a system are largely hidden from view and must be interrogated through field evidence, such as landforms, animals and plants. The recognition of this evidence by a student is important, so that a significant part of this book describes and illustrates a range of coastal landforms and ecosystems. At present, however, one of the most exciting aspects of coastal studies is system dynamics, investigating the response of the physical environment to external forces, such as wave-energy, sea-level change and human interference. In exploring dynamics here, one will hopefully gain an intimate understanding of the interrelationships within the coastal system; therefore, much of the text is devoted to this.

The diversity of coastal environments is a major attraction for their study. An introductory text like this, however, cannot hope to cover this variety, either in breadth or depth. Therefore, the book is divided based on the dominant processes operating on a given coastline, be it waves, tides or rivers. The role of sea level is also covered, and with its increased prominence in the media since the late 1980s, as a consequence of global warming, relatively detailed explanations are given of how sea-level curves are produced and their accuracy, and the mechanisms of sea-level rise under climate change and its management. Further discussion of management issues is the subject of the final chapter. However, case studies of human impact on the coast and associated management issues are scattered throughout the text where appropriate.

Finally, in recent years there has been a growing appreciation of the variation in the rate of coastal change. For many years it was held that coasts developed more or less exclusively through the gradual (high frequency/low magnitude) action of waves, tides and wind. However, many scientists now suggest that low frequency/high magnitude events, such as storms and tsunami, may play dominant roles in the evolution of some coasts, as has been seen in recent years. I have attempted to provide an unbiased approach, incorporating both of these opposing views throughout the text.

1 Coastal systems: definitions, energy and classification

The space occupied by the coast is not easily defined. It is a complex environment that has attributes belonging to both terrestrial and marine environments, which defies a truly integrated classification. This chapter covers:

- the definition of the coast from scientific, planning and management standpoints
- the sources of energy that drive coastal processes
- the architecture and working of coastal systems, introducing concepts of equilibrium and feedbacks
- an introduction to coastal classifications, with an emphasis on broad-scale geological and tectonic controls
- a discussion of the complexities of terminology used in studying coastal systems

1.1 Introduction

1.1.1 Defining the coast

The coast is simply where the land meets the sea. However, applying this statement in the real world is not that straightforward. It is not always easy, for instance, to define exactly where the land finishes and the sea begins. This is particularly so for extensive low-lying coastal wetlands, which for most of the time may be exposed and apparently terrestrial, but a number of times a year become submerged below high tides – does this environment belong to the sea or to the land, and where should the boundary between the two be drawn? It is much more meaningful, therefore, not to talk of coastlines, but of **coastal zones**, a spatial zone between the sea and the land. Usefully, this has been defined as the area between the landward limit of marine influence and the seaward limit of terrestrial influence (Carter, 1988). If we accept this definition, then coasts often become wide spatial areas, for example, encompassing land receiving sea-spray and blown sand from beach sources, and out to sea as far as river water penetrates, issued from estuaries and deltas.

Management Box 1.1

Definitions of the coastal zone for planning and management

The definition of the coastal zone given in section 1.1.1 is very much for the use of physical scientists studying the coast. However, for planning and management purposes, where administration is involved, the coastal zone is much more variably defined. Kay and Alder (2005) give a range of definitions used by various organisations in international and national government. Some definitions are known as distance definitions, whether fixed or variable, where the coastal zone is defined as being so many kilometres landward, and so many nautical miles seaward, of the shoreline. Other definitions do attempt to recognise and incorporate aspects of the working complexity of the coastal zone. In abbreviated form, these include:

- 'the coastal waters and the adjacent shorelands strongly influenced by each other, and includes islands, transitional and intertidal areas, salt marshes, wetlands and beaches. The zone extends inland from the shorelines only to the extent necessary to control shorelands, the uses of which have a direct and significant impact on the coastal waters' (United States Federal Coastal Zone Management Act)
- 'as far inland and as far seaward as necessary to achieve the Coastal Policy objectives, with a primary focus on the land-sea interface' (Australian Commonwealth Coastal Policy)
- 'definitions may vary from area to area and from issue to issue, and that a pragmatic approach must therefore be taken' (United Kingdom Government Environment Committee)
- 'the special area, endowed with special characteristics, of which the boundaries are often determined by the special problems to be tackled' (World Bank Environment Department).

1.1.2 Coastal energy sources

Coasts are not static environments and are in fact highly dynamic, with erosion, sediment transport and deposition all contributing to the continuous physical change that characterises the coast. Such dynamism requires energy to drive the coastal processes that bring about physical change, and all coasts are the product of a combination of two main categories of processes driven by different energy sources (Fig. 1.1):

1. The first category of processes is known as the **endogenetic processes**, so-called because their origin is from within the earth. Endogenetic processes are driven by geothermal energy which emanates from the earth's interior as a product of the general cooling of the earth from its originally hot state, and from radioactive material, which produces heat when it decays. The flux of geothermal energy from the earth's interior to the surface is responsible for driving continental drift and is the energy source in the plate tectonics theory. Its influence on the earth's surface, and the coast is no exception, is to generally raise **relief**, which is to generally elevate the land.

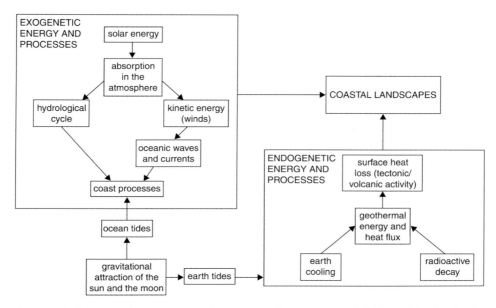

Figure 1.1 Endogenetic and exogenetic energy and processes and their contribution to the development of coastal landscapes.

2. The second category of processes is known as **exogenetic processes**, which are those processes that operate at the earth's surface. These processes are driven by solar energy. Solar radiation heats the earth's surface which creates wind, which in turn creates waves. It also drives the **hydrological cycle**, which is a major cycle in the evolution of all landscapes, and describes the transfer of water between natural stores, such as the ocean. It is in the transfer of this water that rain falls and rivers flow, producing important coastal environments, such as estuaries and deltas. The general effect of exogenetic processes is to erode the land, such as erosion by wind, waves and running water, and so these processes generally reduce relief (however, sand dunes are an exception to this rule, being built up by exogenetic processes).

A third source of energy that is important for coasts is that produced by gravitational effects of the moon and sun. Principally such gravitational attraction creates the well-known ocean tides which work in association with exogenetic processes, but they also produce the lesser-known earth tides which operate in the molten interior of the earth and assist the endogenetic processes.

Ultimately, all coastal landscapes are the product of the interaction of these broad-scale process categories, so where endogenetic processes dominate, mountainous coasts are often produced, whereas many coastal lowlands are dominated by exogenetic processes. Commonly, however, there is a more subtle balance between the two, with features attributable to both process categories present.

1.2 Coastal systems

Natural environments have for some time been viewed as systems with identifiable inputs and outputs of energy (a **closed system**) or both energy and material (an **open system**), and where all components within the system are interrelated (Briggs *et al.*, 1997). The boundaries of a system are not always easily defined, as we discovered in section 1.1.1 when trying to define the coast. Where we can identify a relationship between inputs and outputs, but do not really know how the system works, then we are dealing with a **black box system** (Fig. 1.2); the coast as a whole may be viewed as a black box system. A study of the system may reveal a number of subsystems within it, linked by flows of energy and matter, known as a **grey box system**; a coastal example of this may be a cliff system being eroded by wave-energy, which then supplies an adjacent beach system with sediment. Further investigation may reveal the working components of the system, with energy and material pathways and storages, known as a **white box system**; following on from our previous example, these components may include the rock type that the cliff is composed of, the type of erosion operating on the cliff, sediment transport from the cliff to the beach, beach deposition and its resulting morphology.

1.2.1   System approaches

At the finest scale then, a system comprises components that are linked by energy and material flows. However, there are four different ways in which we can look at physical systems.

1. **Morphological systems** – this approach describes systems not in terms of the dynamic relationships between the components, but simply refers to the morphological expression of the relationships. For example, the slope angle of a coastal cliff may be related to rock type, rock structure, cliff height, and so on.

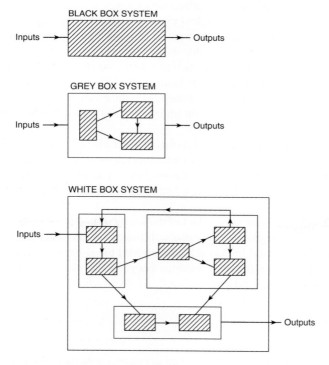

Figure 1.2 Types of systems.

Source: Briggs *et al.* (1997: 5, fig. 1.6).

2. **Cascading systems** – this type of system explicitly refers to the flow or cascade of energy and matter. This is well exemplified by the movement of sediment through the coastal system, perhaps sourced from an eroding cliff, supplied to a beach, and then subsequently blown into coastal sand dunes.

3. **Process-response systems** – this combines both morphological and cascading systems approaches, stating that morphology is a product of the processes operating in the system. These processes are themselves driven by energy and matter, and this is perhaps the most meaningful way to deal with coastal systems. A good example is the retreat of coastal cliffs through erosion by waves. Very simply, if wave-energy increases, erosion processes will often be more effective and the cliff retreats faster. It is very clear from this example, that the operation of a process stimulates a morphological response.

4. **Ecosystems** – this approach refers to the interaction of plants and animals with the physical environment, and is very important in coastal studies. For example, grasses growing on sand dunes enhance the deposition of wind-blown sand, which in turn builds up the dunes, creating further favourable habitats for the dune biological community, and indeed may lead to habitat succession.

1.2.2   The concept of equilibrium

Coasts are dynamic and they change frequently. These changes are principally caused by changes in energy conditions, such as wave-energy for example, which may increase during storms. The morphology of the coast responds to changes in energy because it aims to exist in a state of equilibrium with the reigning processes (Smithson *et al.*, 2002). However, there are three types of equilibrium (Fig. 1.3):

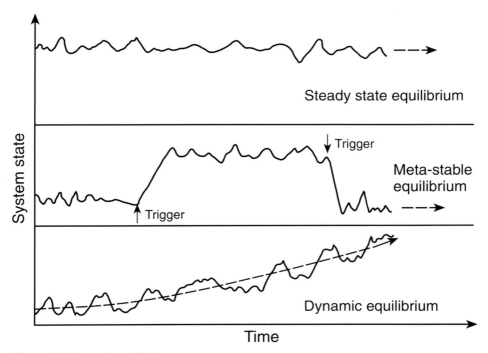

Figure 1.3 Types of equilibrium.

Source: Briggs *et al.* (1997: 7, fig. 1.10).

1. **Steady-state equilibrium** – this refers to a situation where variations in energy and the morphological repsonse do not deviate too far from the long-term average. For example, along a coast that experiences relatively consistent wave-energy conditions, the gradient of a beach may be steeper at certain times of the year, and shallower at others, but the average annual gradient is similar from year to year.
2. **Meta-stable equilibrium** – this exists where an environment switches between two or more states of equilibrium, with the switch stimulated by some sort of trigger. An example of this includes the action of high-energy events, such as storms or tsunami, which can very rapidly switch a coastal system from one state of equilibrium to another, by removing or supplying large volumes of beach sediment for example. Also, human activity often has this effect on coastal environments.
3. **Dynamic equilibrium** – like meta-stable equilibrium, this too involves a change in equilibrium conditions, but in a much more gradual manner. A good example is the response of coasts to the gradual rise in sea levels that we have experienced through the twentieth century as a result of climate change.

1.2.3 System feedbacks

Understanding states of equilibrium in a system requires some knowledge of feedbacks within a system. Feedbacks occur as the result of change in a system and they may be either positive or negative, respectively switching the system to a new state of equilibrium or attempting to recover to the system's original state of equilibrium. **Positive feedbacks** therefore tend to amplify the initial change in the system so that, for example, the ridge of a coastal sand dune breached by storm wave erosion may be subsequently laterally undercut by wind erosion, so fragmenting the dune ridge and leaving it more susceptible to further wave erosion (Fig. 1.4a). Ultimately, the entire dune ridge may be relocated further inland and a new state of equilibrium reached. **Negative feedbacks**, however, tend to dampen the effect of the change. For example, sand eroded during a storm from the front of sand dunes at the back of a beach may be redeposited as offshore sand bars, which help to protect the beach-dune system from further storm waves, by reducing the amount of wave energy reaching the dune front (Fig. 1.4b). Managing coastal systems requires a detailed knowledge of feedbacks, as all too frequently, as we shall see throughout this book, human intervention in one part of a coastal system often leads to a number of apparently unforeseen and undesirable feedbacks.

1.3 The classification of coasts

Because there is such a wide variety of factors that affect coasts, it has been very difficult to actually create an integrated classification scheme (Finkl, 2004). As a result there have been a number of attempts to classify coasts according to single parameters, such as wave or tidal environment, geology, and tectonic setting. Waves and tides are covered individually in other chapters in this book (see Figs 2.1, 3.4 and 3.6), and furthermore are usually only applicable on the local to regional scale. Here we will concentrate on the broad-scale geological and tectonic settings of coasts, which often are applicable to coasts along entire continental margins.

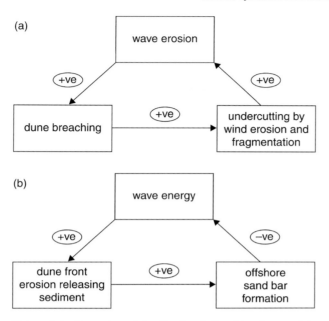

Figure 1.4 Examples of feedback relationships in coastal systems. (a) Example of positive feedback where a positive relationship exists between wave erosion and dune breaching (meaning that an increase/decrease in one, i.e. wave erosion, increases/decreases the other, i.e. the amount of dune breaching), which in turn has a positive relationship to subsequent dune undercutting through wind erosion, so further fragmenting the dune ridge and leaving it more susceptible to further wave erosion. Postive feedback involves either no negative relationships, as in this example, or an even number of them. (b) Example of negative feedback. Note that a negative relationship, such as here between offshore sand bar formation and wave-energy, means that as one increases/decreases the other decreases/increases in response. Negative feedbacks always involves an odd number of negative relationships.

1.3.1 Geological classification

E. Suess in 1888 put forward a coastal classification based on geological structure and its orientation as regards the general trend of the coastline. On this premise he recognised two types of coasts:

1. **Pacific type** – the orientation of the rock structure lies parallel to the coastline. This type of coast is also known as accordant or concordant, and often forms rather straight coastlines interrupted by relatively small embayments. The Dalmatian coast of the former Yugoslavia, in the Mediterranean, is an excellent example of a Pacific type coast.
2. **Atlantic type** – the orientation of the rock structure is at right angles (perpendicular) to the coastline. This coastal type is often known as discordant, and is characterised by prominent headlands and embayments. The southwest coast of Ireland, including Bantry Bay, is a good example of this coastal type.

Horsfall (1993) explores the application of this classification to the classic coastal scenery of Dorset, England. Although it is useful, this scheme is limited in that it gives no indication as to the dynamics of a coastline: is it submerging/emerging, or is deposition/erosion dominant? It is just a statement of the relationship between rock structure and general coastline orientation, which is a random relationship. Small-scale structure can also influence the character of a coast, such as the attitude of joints in granite (Plate 1.1).

Plate 1.1 Small-scale rock structure influences the character of a coastline. Here along the granite coast of western Brittany (France), two contrasting coasts exist: (a) at Concarneau the intertidal zone has a very smooth appearance because the joints in the granite are quasi-horizontal, whereas (b) at Pointe Karreg Leon, near Audierne, the intertidal zone is very jagged because the jointing is quasi-vertical.

Scientific Box 1.2

Integrating the geological classification

Bishop and Cowell (1997) have investigated the relationship between rock structure, river drainage patterns, sea-level change and sedimentology along the eastern Australian coastline. They confirm that Pacific type coasts tend to possess abundant small embayments, whereas embayments are larger on other coasts. However, their analysis suggests that in general, embayment size is related to the size of the river draining into the embayment rather than geological control. They also suggest that sea level has an influence on coastal morphology here, with higher sea levels resulting in a more crenulate and compartmentalised coast with smaller embayments, and lower sea levels creating longer embayments and a more open, less compartmentalised coast. This study does much to try and integrate geology into a more meaningful understanding of coastal development.

1.3.2   Tectonic classification

The theory of plate tectonics describes the creation and destruction of crustal material, and in doing so it explains the movement of the continents around the globe. The earth's surface comprises a number of continental and oceanic crustal plates (Fig. 1.5). Each one

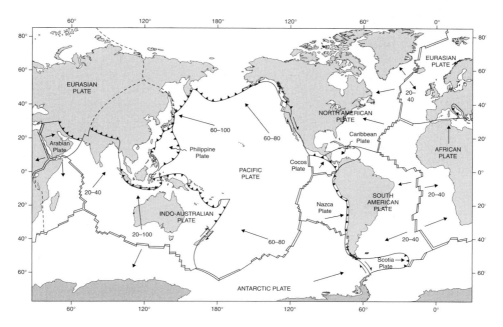

Figure 1.5 The distribution of the earth's crustal plates and their various boundary types. Parallel lines = constructive (divergent) boundary; toothed lines = destructive (convergent) boundary (teeth pointing toward subduction); and single/broken lines = conservative (transform) boundary. Arrows and figures indicate plate movement in mm/year.

Source: Briggs *et al.* (1997: 36, fig 3.5).

of these plates is bounded by zones where either new crust is created (constructive boundary) or where old crust is destroyed (destructive boundary). Plate movement is usually away from constructive boundaries (divergence), and towards destructive boundaries (convergence). At the broad scale, there is often a consistent relationship between the characteristics of a coastline and the type of plate boundary that it is nearest to. Inman and Nordstrom (1971) recognise a number of coastal types based on the plate boundary they are associated with. Davis (1997) reviews **leading edge coasts**, **trailing edge coasts** and **marginal sea coasts**, all of which are briefly described below.

1.3.2.1 Leading edge coasts

These occur where a continental plate converges with an oceanic plate at a destructive boundary. Because of this, these types of coasts are also known as convergent margin coasts. Continental crust is less dense and, therefore, more buoyant than oceanic crust, resulting in the oceanic plate going under or subducting beneath the continental plate. The compressional forces created by convergence cause the rocks along the coast to buckle, fold and fault, uplifting them to create chains of coastal mountains. Earthquakes are commonly associated with this coastal type (see Box 1.3). The continental shelf in front of the mountain chain is usually narrow or even absent, and therefore the gradient from the top of the coastal mountains to the sea floor is usually very steep. Therefore, although much sediment is eroded from the uplifting coastal mountains by fast-running streams, it is usually lost into deep water via submarine canyons when it is introduced into the sea. Commonly then, leading edge coasts are characteristically mountainous, dominated by erosional processes, and so often rocky. The longest leading edge coast occurs more or less along the entire western seaboard of the Americas (Plate 1.2), both northern and southern continents, with the exception of southern California.

Plate 1.2 The rugged coast of Oregon (USA), looking north toward Hecata Head lighthouse on Highway 101, is a good example of a leading edge coast being dominated by erosion with very limited coastal deposits.

Scientific Box 1.3

Tectonic compression, earthquakes and leading edge coasts

Leading edge coasts are subject to compressional tectonic forces produced as the two plates converge. Plates do not move past each other continuously, but through sudden movements known as seismic events, such as an earthquake. Earthquakes represent the release of the energy built up by the frictional strain of the two plates. During eathquakes the coast often sinks, sometimes by over 1 m, a phenomenon known as **coseismic subsidence**. In the period between earthquakes compressional energy builds up again, referred to as **interseismic strain accumulation**, which has the effect of raising the land back out of the sea. Therefore, leading edge coasts often go through a cycle of rapid submergence during an earthquake, followed by a period of gradual elevation. This process is often dramatically seen with the drowning of coastal forests that colonised the coast during the interseismic period, which may last several hundreds of years. Leonard *et al.* (2004) provide an example from the Washington-Oregon coast, USA.

1.3.2.2 Trailing edge coasts

These are coasts that form as plates rift apart due to divergence, allowing the ocean to inundate the rift to create a new sea. Once formed, these coasts are carried away from the diverging boundary as passive continental margins (Plate 1.3). Shortly after rifting, the coasts comprise relatively high relief and possess a fairly steep gradient, and in many respects their topography is very similar to leading edge coasts. These are known as **neo-trailing edge coasts** and the present-day coasts of the Red Sea belong to this subdivision. As divergence progresses the sea expands and erosion of the coast increases, both by wave activity at the shoreline and through the action of high-energy streams flowing down steep hills. In this way, the continental shelf begins to widen, the relief is lowered, and the **afro-trailing edge coast** is created. As the name implies, most of the African coastline is of this type, with the exception of course of some Mediterranean and Red Sea coasts. Africa has been tectonically stable for many millions of years, and although the continental shelf is relatively wide now, large sedimentary features such as deltas are limited in comparison with the most mature trailing edge coastal subtype, the **amero-trailing edge coast**. This mature coastal type is characteristic of the eastern seaboard of the Americas, with an extensive coastal plain existing along the North American section, and vast deltas, such as the Amazon Delta, characterising the South American section.

1.3.2.3 Marginal sea coasts

This is a relatively uncommon coastal type and occurs where plate convergence takes place offshore, with a relatively wide continental shelf separating the plate boundary from the coast. In many ways, this coastal type is similar to trailing edge coasts, especially in the sedimentary features that may develop, but it suffers earthquakes more regularly and may experience limited tectonic movement. The Gulf of Mexico coasts are an example, with the volcanic islands of the Caribbean marking the zone of plate convergence.

Plate 1.3 The trailing edge coast of eastern Australia (this section of the coast is to the north of Cairns, Queensland). Although at first glance it is similar to a leading edge coast, the continental shelf is wider here, supporting more extensive coastal deposits and, indeed, the Great Barrier Reef situated offshore.

1.4 Coastal terminology

As with all scientific disciplines, the study of coasts has generated a vast lexicon of terms to describe landforms, processes, deposits, habitats and ecosystems. Hopefully readers will agree that one of the most valuable parts of this book is the fairly extensive glossary. I have included all the important coastal terms that one will come across in the literature, in the hope that students will not be discouraged by the sometimes bewildering array of terms that are used, not only to describe different attributes of coastal systems, but sometimes apparently the same attributes. Rarely, however, are these terms precisely synonymous, and important, if subtle, differences in meaning do exist. A command of coastal terminology serves to better understand coastal systems, and also to enhance the communication of the subject. I therefore encourage you not to pass over unfamiliar terms, but to try and grasp their meaning from both the text in which it occurs, and with reference to the glossary. For when one goes beyond this introductory text to delve into the primary literature in scientific journals, which I thoroughly recommend, one quickly discovers that no allowance is given to readers with a limited vocabulary on the subject, and the value of spending time reading such articles is correspondingly diminished.

A useful example of similar yet subtly different terms involves the standard subdivision of the coastal zone (Fig. 1.6). Two sets of terms are shown in Fig. 1.6; however, one subdivides the coastal zone based on morphological changes (**backshore**, **foreshore**, **inshore** and **offshore**), whilst the other is based on the types of wave processes that operate in different parts of the coastal zone (**swash zone**, **surf zone** and **breaker zone**, which together constitute the **nearshore zone**). The same space may also be subdivided on sedimentological grounds, according to the type of sediment, sedimentary structures and/or depositional processes that occur (Reading and Collinson, 1996):

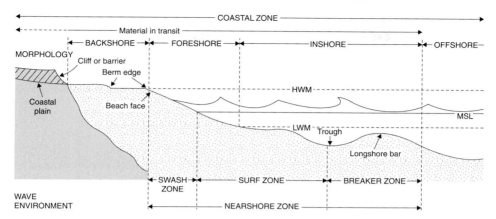

Figure 1.6 Subdivision of the coastal zone using both morphological and wave process terminology.

Source: Briggs *et al.* (1997, Fig. 17.1).

- the **beach zone** occurs between mean low water and the landward limit of wave activity and mainly possesses parallel-bedded sand layers;
- the **shoreface zone** occurs between mean low water and the mean fairweather wave-base (see sections 2.2 and 2.6.1 for explanation) and is characterised by wave-swept sand ripples;
- the **offshore transition zone** lies between the mean fairweather wave-base and the mean storm wave-base (again, see sections 2.2 and 2.6.1) and is dominated by sediment deposition during storms;
- the **offshore zone** lies below the mean storm wave-base and is characterised by fine-grained sediment settling out of the water.

Summary points

- The coastal zone may be defined as the area between the landward limit of marine influence and the seaward limit of terrestrial influence.
- Coasts, like all landscapes, are the product of the combined influence of endogenetic and exogenetic energy and processes.
- Coasts operate as open systems, with the flow of energy and material into, through and out of the coastal system. Coastal morphology aims to attain a state of equilibrium with the reigning processes; however, change within the system may trigger feedbacks, which may either accentuate the change (positive feedback) or help to minimise its effect (negative feedback).
- The classification of coasts is problematic, but on the broad scale the use of tectonic conditions helps us subdivide the global coastline into leading edge, trailing edge and marginal sea coasts, each possessing a range of characteristic attributes.
- The use of diverse and sometimes confusing terminology in studying coasts is important for the student to command.

Discussion questions

1. Examine the problems in constructing a meaningful definition of the coast.
2. The study of coasts is well suited to a systems approach. Discuss why this might be so.
3. Assess the significance of tectonics in the development of a global coastal classification scheme.

Further reading

See also

Administrative coastal management issues, section 6.2.4
The development of shore platforms, section 2.3.2
Wave classification of coasts, section 2.1
Tidal classifications of coasts, sections 3.2 and 3.3
Wave and beach terminology, sections 2.2 and 2.6.1

Introductory reading

An Introduction to Physical Geography and the Environment. J. Holden (editor). 2004. Prentice Hall, Harlow, 696pp.
Chapter 17 *Coasts* provides a useful introductory overview.

Fundamentals of Geomorphology (2nd edn). R. J. Huggett. 2007. Routledge, London, 458pp.
Part II deals with the general influence of geological structure on landscape development, whilst Chapter 11 specifically examines coastal landscapes.

Fundamentals of the Physical Environment (3rd edn). P. Smithson, K. Addison and K. Atkinson. 2002. Routledge, London, 627pp.
Chapters 1, 2 and 17 are all useful in expanding further the themes introduced in this chapter.

Introducing Physical Geography (4th edn). A. H. Strahler and A. Strahler. 2005. Wiley, New York, 752pp.
Chapter 17 *Landforms and Rock Structure* is useful for understanding the relationship between earth processes and landform development.

Physical Geography: A Landscape Appreciation. T. L. McKnight and D. Hess. 2004. Prentice Hall, Harlow, 640pp.
Chapters 13 and 20 outline aspects of landform development and coastal landscapes respectively.

Advanced reading

Beaches and Coasts. R. A. Davis Jr and D. M. Fitzgerald. 2004. Blackwell Science, Oxford, 419pp.
A comprehensive treatment of coastal geology and geomorphology.

Coastal Dynamics and Landforms. A. S. Trenhaile. 1997. Oxford University Press, Oxford, 365pp.
An advanced-level text dealing with process and form at the coast.

Coastal Geomorphology: An Introduction. E. Bird. 2000. Wiley, Chichester, 322pp.
An extensive overview of coastal processes and geomorphology.

Coasts: Form, Process and Evolution. C. D. Woodroffe. 2002. Cambridge University Press, Cambridge, 623pp.
An encyclopaedic treatment of coasts for advanced undergraduate and graduate students.

Introduction to Coastal Processes and Geomorphology. G. Masselink and M. G. Hughes. 2003. Arnold, London, 354pp.
An accessible advanced introduction to coastal processes and geomorphology.

The Evolving Coast. R. A. Davis Jr. 1997. Scientific American Library, New York, 233pp.
A concise introduction to the physical development of coastlines.

Dynamics of Coastal Systems. J. Dronkers. 2005. World Scientific, Singapore, 519pp.
An advanced text in ocean engineering aspects of coastal systems.

The periodicals *Journal of Coastal Research*, *Estuarine, Coastal and Shelf Science*, *Shore and Beach*, *Journal of Coastal Conservation*, *Marine Geology* and others should be consulted regularly for individual research papers, themed sections, and special issues devoted to particular topics.

2 Wave-dominated coastal systems

All coasts are affected to a certain degree by wave activity, and waves provide the energy that drives many of the coastal processes that create many of the worlds most spectacular coasts. This chapter covers:

- **the origin and characteristics of waves and their processes**
- **the dynamic interaction between waves and coastal systems**
- **the geomorphology and ecology of erosional and depositional wave-dominated coastlines**
- **examples of environmental, management and engineering issues affecting wave-dominated coastlines**

2.1 Introduction

Wave-dominated coasts are amongst the most familiar of all coastal environments. Sandy beaches are popular holiday destinations and dramatic cliffed coastlines are often frequented by walkers, hikers, rock-climbers and hang-gliders. They are very dynamic and sensitive systems, often in equilibrium in the natural state, but susceptible to even slight interference from human activity.

On a global scale, coastal wave environments reflect to a large degree the climatic conditions experienced in a given region, such as wind speed and duration, as well as the size of the ocean concerned (Fig. 2.1). For example, the mid-latitudes in both northern and southern hemispheres experience regular storms which in turn can produce very large waves. Storm-wave-affected coasts, therefore, occur in the path of major storm tracks, such as northwest North America from Oregon to Alaska, northeast North America from Florida to Newfoundland, northwest European coasts of France, British Isles and Scandinavia, Atlantic and Pacific coasts of southern South America, and southerly coasts of Australia and New Zealand. Storm-wave-affected coasts also occur in tropical latitudes, such as southeast Asia and the Arabian Sea, where they regularly experience tropical cyclones. Outside these storm belts, waves on open ocean coasts tend to be smaller, but may still be significant in influencing coastal development. This is at a minimum, however, in the subtropical doldrums, where wind activity is least intense. Enclosed sea coasts, such as the Mediterranean and Red Seas, experience minimum wave conditions due to their size, which prohibits the development of large waves (see below). However, it is coasts in polar regions that are least affected by waves because for long periods of the year (if not permanently) their coasts are protected by the formation of coastal sea ice.

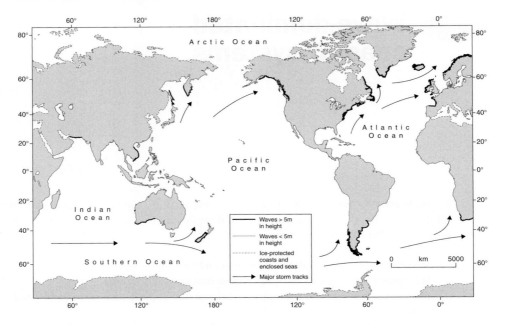

Figure 2.1 The distribution of global wave environments: coasts experiencing >5 m high waves are storm-affected.

Source: Briggs *et al.* (1997).

2.2 Waves

All waves possess alternating high crests and low troughs, and can be described (Fig. 2.2) with reference to

- **wave-height (H)**, the vertical distance between the wave trough and crest
- **wave-length (λ)**, the horizontal distance between consecutive crests or troughs
- **wave-period (T)**, the time interval between consecutive wave crests or troughs passing a fixed point.

Waves represent the transfer of energy, behaving differently depending on whether they are in deep or shallow water (Fig. 2.3). In deep water, there is only minor movement of water in the wave. Although energy passes forward within a wave, water follows a circular pattern. A water particle may, from a starting position at the bottom of a wave trough, move up the face of an oncoming wave, against the direction in which the wave is travelling, to reach the crest and return to its original position. It will then travel down the back of the wave under gravity, this time moving forward with the wave, but finishing back at the bottom of the next wave trough at its original starting position. This closed loop, where a water particle moves in a circular orbit, has given rise to the name **oscillatory wave**. The diameter of water particle orbits decreases with depth through the water column, so that eventually a depth is reached below which the water is unaffected by waves passing overhead. This depth is called the **wave-base**, and helps to define deep- and shallow-water waves, as deep-water waves occur in water deeper than their wave-base, whilst shallow-water waves are travelling through a water depth that is less than their wave-base. Wave-

Figure 2.2
(a) Wave description: wave-length (λ) is the horizontal distance (m) between consecutive wave crests or troughs, and wave-height (H) is the vertical distance between trough and crest. (b) The circular orbit of an oscillatory wave.

Source: Park (1997).

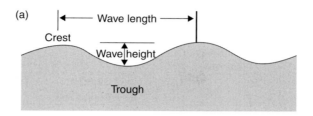

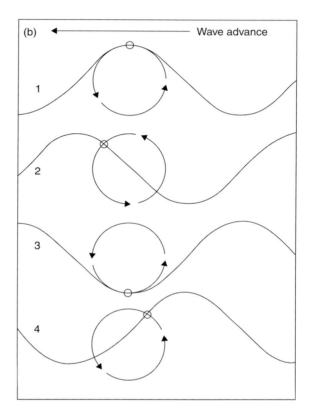

base can be established with reference to wave-length, and various authors have defined it as a water depth equal to a range between a half and a quarter of the wave-length. Pethick (1984) employs the latter in defining wave-base, although he does state that some water movement can take place below this depth. The significance of wave-base cannot be overstated, as waves are perhaps the most important geomorphological agent on the coast. Waves can erode the coast, transport sediment, and ultimately deposit sediment, but all this activity takes place at and above wave-base. So in determining wave-base, we are going some way to determining the spatial extent of wave influence on a coast.

The study of waves involves some quite sophisticated mathematics, and although I will try to avoid most equations, there are a few which are very useful to have to hand. The first of them introduced below helps us to establish the wave-length of deep-water waves, for although the definition of wave-length is straightforward, its actual measurement at sea is an extremely difficult task. However, wave-length is related to wave-period, a wave characteristic that is far easier to determine. All that is required for an observer to establish wave-period is a fixed reference point and a stopwatch, recording the time taken for two

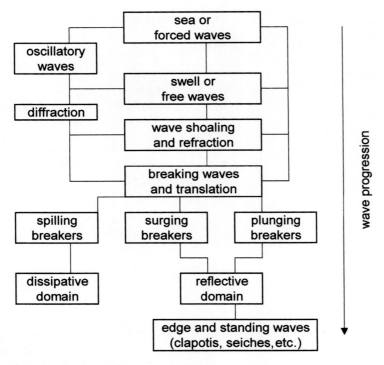

Figure 2.3 Wave progression through deep- to shallow-water processes.

wave crests to pass the fixed point. Wave-length can then be calculated using the following equation:

$$\lambda_{\mathrm{o}} = \frac{gT^2}{2\pi}$$

where λ_{o} is the wave-length of a deep-water wave, g is acceleration due to gravity (a constant at 9.81 m/s²), T is wave-period in seconds, and π is the ratio of the circumference of a circle to its diameter, approximately 3.142. Of course, the significance of this equation lies in our ability to now calculate the all-important wave-base from the wave-length. It also indicates that only modest increases in wave-period result in quite substantial increases in wave-length (Pethick, 1984). Another useful equation, which again incorporates wave-period values, determines wave-velocity or celerity (C):

$$C_{\mathrm{o}} = \frac{gT}{2\pi}$$

where C_{o} is wave-velocity (m/s) of a deep-water wave. An observation that can be made from this equation is that waves with a high wave-period, and therefore a long wave-length, travel faster than low wave-period, short wave-length waves (Suhayda and Pettigrew, 1977; Pethick, 1984).

Because we are dealing with coasts, all waves that are relevant to our study eventually encounter water depths shallower than wave-base (a depth equal to a quarter of the wave-length). When this happens the sea-bed begins to interfere with the oscillatory motion of the deep-water wave, and the wave starts **shoaling**, so becoming a shallow-water wave (Fig. 2.4). Wave-length and wave-velocity both decrease, but the energy from these reductions contributes to an increase in wave-height, which in turn leads to steepening of the wave-front. The circular orbits of the deep-water wave become progressively elliptical, with the long axis of the ellipse parallel with the sea-floor, promoting alternating onshore-offshore currents that may be capable of picking up and transporting grains of sediment. The wave is now able to transmit matter as well as energy, and becomes a **translatory wave**. A shallow-water wave will usually break when it encounters water depth that is less than wave-height (the **break-point**). In such shallow water, equivalent to less than one-twentieth of the wave-length, the equations for calculating deep-water wave-length and wave-velocity can no longer be used, and are substituted by:

$$\lambda = T\sqrt{gd}$$

and

$$C = \sqrt{gd}$$

respectively, where d is depth (m).

Where an **incident wave** encounters wave-base along only part of the wave-front, the wave will undergo **refraction**. This situation may arise, for example, where submarine topography is variable, or where an incident wave approaches the shore obliquely with submarine topography descending uniformly offshore. Refraction is an important wave modification process because it affects the distribution of wave-energy along the shore (Fig. 2.5). Shoaling promotes deceleration of wave-celerity, so that those parts of a wave in water shallower than wave-base will slow, whilst sections of the wave still in water deeper than wave-base will maintain their celerity. In this way the wave will bend or refract. On a map of refracting wave-fronts it is possible to draw lines called **orthogonals** normal to the wave-front and extending to the shore. Where orthogonals converge on a shore, wave-energy is concentrated, whilst where orthogonals diverge on a shore, wave-energy is dissipated. Therefore, along relatively straight shorelines with parallel and gradually descending submarine contours, there will be very little wave refraction and wave-energy will be

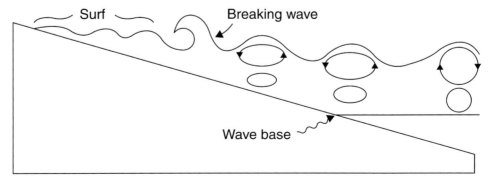

Figure 2.4 Wave shoaling as an oscillatory wave enters shallow water with a transformation of water particles from a circular to an elliptical orbit, leading to wave breaking.

Source: modified from Briggs *et al.* (1997).

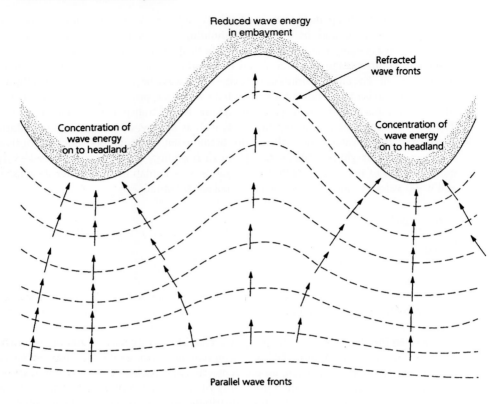

Figure 2.5 Wave refraction on an irregular coastline: the arrows are orthogonals drawn at right-angles to the wave-fronts (dashed lines); energy is concentrated where orthogonals converge and dissipated where they diverge.

Source: from French (1997).

uniformly distributed on the shore. However, along a highly indented coastline, characterised by headlands and embayments, wave-base is likely to be encountered off the headlands first. This would result in refraction of the wave around the headlands, convergence of the orthogonals and concentration of wave-energy on the headlands, and a corresponding divergence of orthogonals and dissipation of wave-energy in the bays. In this way, erosion may be expected to dominate on the headlands, whilst depositional processes are more likely to characterise the lower energy embayments (McCullagh, 1978). Because wave-length and, therefore, wave-base of wind-created waves varies with wind strength (Young, 1999), patterns of refraction may vary seasonally along a given coast.

Diffraction is another important wave modification process. When an incident wave encounters an obstacle in its path, such as an island or harbour wall, a wave shadow zone is created in the lee of the obstacle. However, as the wave passes the obstacle it is able to propagate or diffract into the shadow zone. It is able to do this because as a passing wave crest falls due to gravity, water is no longer confined by a continuous wave-front and so is allowed to escape or 'bleed' laterally into the shadow zone, creating waves that are smaller than the parent wave (Stowe, 1996). Therefore, apparently protected shadow zone shores are still susceptible to wave activity and breaking.

The manner in which waves break against a shore is determined by the steepness of the incident wave, water depth and also the gradient of the shore (Galvin, 1968; Summerfield, 1991).

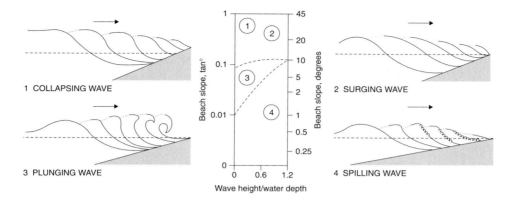

Figure 2.6 Types of breaking waves as a function of wave-height, water depth and beach slope gradient.

Source: from Briggs *et al.* (1997).

- **Spilling breakers** tend to characterise shores with a low-angled gradient, regardless of wave steepness, where water from the wave-crest spills or cascades down the wave-front producing lots of foam.
- **Plunging breakers** occur on steeply inclined shores and are steep-fronted waves that tend to curl over and crash down on the shore, again producing lots of foam.
- **Surging breakers** are low waves that fail to break fully, the crest collapses, and the base of the wave-front advances up the shore as a rather foamless breaker.

Of course, these breakers occur along a continuous spectrum of breaker types, and intermediate breakers have been described (e.g. collapsing breakers) (Fig. 2.6).

The type of breaker is also associated with the way in which wave-energy is expended. Spilling breakers often roll in over relatively wide low-angled shores, upon which their wave-energy is dissipated gradually, giving rise to the term **dissipative domain**. Plunging and surging breakers, however, often breaking on steeper shores soon after encountering wave-base, will have a significant proportion of their energy reflected back out to sea as reflected waves, and hence operate in a **reflective domain**. Reflection of wave-energy may also occur off hard structures, whether natural (e.g. cliffs) or artificial (e.g. harbour walls). Reflected waves often travel at right angles to the shore as **edge waves** and interact with incoming incident waves. Where the crests of both edge and incident waves intersect, the heights of the waves combine to increase the wave-height at the intersection, and intersecting troughs combine to lower trough depth, resulting in both undulating wave-crests and troughs. Where reflected waves approximately equal the characteristics of, and travel in the opposite direction to, incoming waves then a **standing wave** known as a **clapotis** may be produced, in which the water goes up and down, but does not progress. Reflection occurring in a semi-enclosed body of water, such as a cove or harbour, may produce a **seiche**, another type of standing wave, which sloshes back and forth across and out of the water body (Stowe, 1996).

2.2.1   Wind waves

Wind waves are those created by the friction that arises when wind blows over the sea surface (Fig. 2.7). Both air and water are fluids, but with different densities, and as such

Management Box 2.1

Coastal management implications of wave modification processes

As incident waves enter the nearshore zone they become modified by submarine topography, coastal configuration and coastal substrate types, posing special problems for coastal managers and engineers, some of which are listed below.

- **Refraction**. Rises and mounds on the sea-floor may focus wave-energy onto certain stretches of the coast. For example, artificial hardstanding for piers may focus wave-energy onto the coast where the pier is connected to the shore, which in extreme cases may undermine foundations and result in pier detachment from the shore. Also, the famous case of the Long Beach breakwater in California is worth recounting, for in April 1930 it was badly damaged by long wave-length waves from a southern hemisphere storm that were refracted by a 'bump' in the continental shelf as the wave-train passed over it. The wave-energy was so well focused that the coasts neighbouring Long Beach were completely unaffected (Stowe, 1996).
- **Diffraction**. This modification process allows waves to enter into areas perceived to be protected, such as the lee-side of islands, behind breakwaters and within harbours. Island shadow zones are often characterised by choppy wave conditions as waves diffract in from both sides of the island, sometimes creating conditions more hazardous to sailors than exposed coasts, and makes island-hopping quite uncomfortable for those without sea-legs! Also, harbours and breakwaters should be designed to minimise diffraction by aligning structures parallel to the fetch direction, as far as is possible, to protect valuable water-craft and sea-side buildings from storm-wave attack.
- **Reflection**. This modification process transforms incident waves into reflected waves. The formation of reflected waves in confined coves and harbours can create standing waves that may lead to collision and damage of closely moored boats, and produce unsuitable sailing conditions. Therefore, harbour walls should be designed to dissipate and absorb wave-energy rather than reflect it. This is largely achieved by building porous structures, such as rubble-mound barricades and boulder armouring. There is a further advantage to minimising reflection in that reflection can create resonance within the reflecting structure, which may lead to the development of internal weaknesses, degradation and the ultimate replacement of the structure at cost.

frictional drag forms waves at the interface between them. Therefore, wind waves are indirectly created by solar energy and categorised as exogenetic. Many waves are generated by localised storm activity at sea, and whilst the waves are still within the generation area they are termed **sea waves** or **forced waves**. Upon leaving the generation area these waves lose height and energy to become **swell waves** or **free waves**, which may travel many thousands of miles before breaking on a distant shore. This reduction in height and decrease in energy occurs rapidly to begin with, so that an initially 10 m high wave may in its first 200 km of travel from the generation area be reduced to 2 m in height

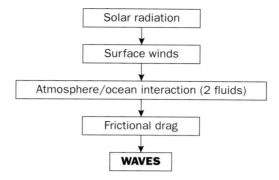

Figure 2.7 The creation of wind waves.

with a corresponding 80–90% energy decrease, but thereafter will diminish only slightly (Komar, 1976; Summerfield, 1991).

The height of a wind-generated wave is closely associated with wind velocity. Observations have shown that wave-height is proportional to the square of wind velocity, so that higher wind velocities create higher waves and generally rougher seas. This relationship can be summarised as:

$$H = 0.031 \times U^2$$

where H is wave-height (m), U is wind velocity (m/s), and 0.031 is an empirically derived constant. Wave-height is also affected by **fetch**, which is the distance over which wind interacts with the sea-surface to create and propagate waves (Plate 2.1). Observations indicate that wave-height is proportional to the square root of the fetch, so that higher waves are associated with a long fetch. This relationship can be summarised as:

$$H = 0.36 \times \sqrt{F}$$

where H is wave-height (m), F is fetch (km), and 0.36 is an empirically derived constant. The combined affect of high wind velocities and a long fetch results in high waves, so that extensive seas like the Atlantic and Pacific Oceans, for example, experience waves in excess of 25 m high (Pethick, 1984; NERC, 1991). When discussing wave-height, however, it is often more meaningful to refer to **significant wave-height (H_s)**, which is the mean of the highest one-third of waves affecting a coast. For example, significant wave-height in the Atlantic Ocean is approximately 2 m.

In a wave-generating area, wind does not create uniform groups of waves, especially during storms, but more often creates a chaotic assemblage of waves with varying wave-lengths and periods. As we discussed above, waves with a long wave-length travel faster than waves with a short wave-length. From this it is clear that swell waves leaving a generating area will eventually sort themselves, with the long wave-length waves in the front and the short wave-length waves at the back of the travelling wave-train. This process is called **dispersion** and is responsible for many Atlantic and Pacific coasts (e.g. southwest Britain, northwest France, California, and New South Wales) receiving long wave-length waves suitable for surfing. In contrast, within enclosed seas, such as the Mediterranean, dispersion is limited by the restricted fetch, resulting in the co-occurrence of long and short wave-length waves that produce choppy wave conditions (Pethick, 1984; Stowe, 1996). Where two groups of waves travelling in the same direction with similar wave-lengths coincide (**superposition**), but perhaps slightly out of phase with each other, they may produce a wave phenomenon known as **surf beat**. As with edge waves discussed earlier, where crests of waves in both wave groups coincide, the heights of the individual waves combine to increase wave-height; however, the combined effect of a wave-trough in one group coinciding with a wave-crest in another group is to reduce wave-height. Therefore, higher than average waves are formed where the two wave groups are in phase, producing ideal surfing conditions, but lower than average heights are achieved where the wave groups are out of phase.

2.2.2　Tsunami waves

Tsunami are waves that are produced by a number of different mechanisms, including displacement of the sea-floor by movement along a fault (associated with an earthquake) or a submarine slide, volcanic activity, and asteroid impact. They have also been called tidal waves and seismic sea waves, but these are inappropriate as tsunami are unrelated to tides and those created by asteroid impacts are cosmogenic rather than seismogenic. There are also meteorological tsunami, which are waves similar in character to seismogenic tsunami but are generated by meteorological conditions, such as squalls and fronts, atmospheric gravity waves and pressure jumps (Montserrat *et al.*, 2006). The word tsunami itself is not ideal, as it is Japanese for harbour wave, and these waves are not confined to harbours. In the case of sea-floor displacement, sea level changes directly above the area of displacement and a tsunami is created. With a submarine slide, the water column subsides to fill the hollow created by the slide, surface water rushes into the area to restore sea level, but overcompensates creating a localised bulge in sea level which subsequently propagates outward as a tsunami (Smith and Dawson, 1990). Asteroid impacts act in the same way as throwing a stone into a pool, where a combination of displacement and energy transfer occurs between the impacting body and the surface water. Displacement may also occur due to slope instability on the flanks of volcanoes that may have generated large tsunami in the past, and present a risk for the future (Whelan and Kelletat, 2003).

In the open ocean, tsunami are similar to wind waves in that wave-heights are often small, commonly no more than 1 m high, and ships at sea would not normally notice a tsunami passing beneath them. However, tsunami possess extremely long wave-lengths, measured in hundreds of kilometres, which means that they may almost always be considered as translatory waves and that they travel at extremely high speeds, for example 600 km/hr in a water depth of 3000 m (Summerfield, 1991). In this way, tsunami are able to rapidly cross large oceans, such as the Pacific, an ocean that is particularly prone to tsunami with at least one tsunami event occurring every 25–50 years in Hawaii. As with all waves, when tsunami enter progressively shallow water, usually as they encounter the continental shelf, their velocity decreases and wave-height increases, often with catastrophic consequences for coastal populations. They may also refract, diffract and reflect, further concentrating their energy on particular parts of the coast, affecting areas in the shadow zones of islands, and creating edge and standing waves, respectively.

Although a tsunamigenic event will usually create not one tsunami, but a series of tsunami waves that may repetitively affect a coastline for up to 24 hours, these events occur only occasionally, even on tsunami-prone coasts. Summerfield (1991) considers tsunami to be of such low frequency that they are of less significance than high frequency wind waves in contributing to the physical development of coasts. However, Bryant (2001, 2008) challenges this paradigm suggesting that along some coastlines, such as New South Wales in Australia, tsunami do make a substantial contribution to coastal evolution, and that coasts

Plate 2.1 The influence of fetch on waves: (a) long fetch waves on the eastern Atlantic coast at Bretignolle-Sur-Mer (western France) characterised by long wave-lengths and high wave-heights; (b) short fetch waves on the northeastern coast of Australia at Yorkey's Knob (near Cairns, Queensland), where the Great Barrier Reef prevents swell waves from the Pacific Ocean entering the coastal waters, thus these waves are created landward of the reef and are characterised by short wave-lengths and low wave-heights.

should be evaluated individually as to the extent tsunami may affect their evolution (e.g. Haslett and Bryant, 2007a). Bryant's (2001) thesis, although controversial, has stimulated renewed interest in the nature of coastal evolution and the role of high-energy events, such as tsunami, that may ultimately lead to a paradigm shift within the discipline. Indeed, Carter and Woodroffe (1994: 15) anticipated, from Bryant and co-workers' early research, that it might lead to the revision of 'many long-held ideas about coastal evolution'. After all, the geomorphological significance of low frequency, but high magnitude events in other physical systems, such as river flood events, has long been appreciated.

Research undertaken in Great Britain, a region not normally associated with tsunami, has suggested that tsunami have affected the coast in the past, and provides a good example of a re-evaluation of the contribution made by such events to coastal evolution. Events include tsunami over 20 m high generated by a massive submarine slide, known as the Storegga Slide, that occurred offshore Norway around 7000 years ago as sea level rose after deglaciation and destabilised sediments accumulated on the continental shelf (Smith *et al.*, 2004). Also, an 8.7 magnitude earthquake occurred along the Azores-Gibraltar Fracture Zone offshore Portugal on 1 November 1755 that resulted in a 3 m high tsunami reaching the shores of southwest Britain and Ireland (Dawson *et al.*, 2000). More controversially, contemporary accounts of a catastrophic flood that occurred in southwest Britain on 30 January 1607 (Fig. 2.8) have been interpreted as describing a tsunami (Bryant and Haslett, 2003; Haslett and Bryant, 2005), which may be associated with tsunami signatures in the coastal landscape (Bryant and Haslett, 2007). Although other authors regard this event as being caused by a storm surge (Horsburgh and Horritt, 2006), the flood, if due to a tsunami, is most likely to have been triggered by an earthquake occurring along an active fault zone offshore southwest Ireland (McGuire, 2005). Other historic British events may also be related to tsunami and initial investigations suggest these events require further examination (Haslett and Bryant, 2007b, 2008).

Figure 2.8 A woodcut depicting the coastal flood of 30 January 1607 in the Bristol Channel and Severn Estuary, UK.

Case Study Box 2.2

Indian Ocean tsunami, 26 December 2004

Around 8am local time, below the eastern Indian Ocean, an earthquake with a magnitude of over 9 caused a massive rupture on the sea-bed that generated a deadly tsunami that fatally struck the coasts of 11 countries around the Indian Ocean perimeter, leaving over 280,000 people dead. Although the tsunami was destructive only within the Indian Ocean, it was pan-oceanic, entering the Pacific and Atlantic oceans (Titov *et al.*, 2005) where it was recorded in Brazil (França and Mesquita, 2007) and even on tide gauges in the United Kingdom (Long and Wilson, 2007). The earthquake was one of the biggest experienced in the last 100 years with an energy release of over 1500 Hiroshima-type atomic bombs. Damage and fatalities caused during the event attracted interest and concern from around the world, with monetary donations exceeding seven billion US dollars.

The physical effects of the tsunami impact along Indian Ocean shores were widespread and varied. The worst affected coast was that of Banda Aceh on the island of Sumatra in Indonesia, where waves attained heights of 24 m along stretches of the coast and up to 30 m high at certain locations (Sumatra International Tsunami Survey Team, 2005). The tsunami inundated about 4 km inland, penetrating coastal lowlands through gaps in barriers (Plate 2.2a) and up estuaries, as well as traversing flat coastal plains. Most houses, with the exception of reinforced concrete buildings, were swept away within a zone 2–3 km from the shore (Tohoku University, 2005). The tsunami caused extensive erosion at the shore, effectively scouring soft-sediment landforms, such as beaches and marshes, and picking up and transporting sediment, including large boulders (Plate 2.2b) and sand, which was deposited as extensive sand sheets decimetres thick. Also, the Banda Aceh coast experienced co-seismic subsidence in places (see Scientific Box 1.3) as indicated by the submergence of tree-clad land surfaces. The combined impacts have generally resulted in coastal retreat, but beaches have reformed locally.

Plate 2.2 continued

Plate 2.2 Impacts of the Indian Ocean tsunami, 26 December 2004. (a) Tsunami inundation at Banda Aceh, Indonesia; the arrows indicate tsunami flow direction as the waves hit the shore; then, where the waves were forced inland past barriers, the flow became focused and wrapped around the back of these barriers. (b) Large boulders carried landward from a rock revetment by tsunami in Banda Aceh, Indonesia.

(a) Photo from website http://walrus.wr.usgs.gov/tsunami/sumatra05/images/FlowDirection.jpg (accessed 20 November 2007). (b) Photo from website http://walrus.wr.usgs.gov/tsunami/sumatra05/Banda_Aceh/0643.html (accessed 20 November 2007). Reproduced with permission of the United States Geological Survey.

The Indian Ocean had no tsunami warning system in place at the time and tsunami were clearly an underrated hazard within the region. However, this 2004 event has brought tsunami into the spotlight, as all coasts are vulnerable, to a lesser or greater degree, to the impact of a destructive tsunami, as argued by Bryant (2001), and they have a major role to play in coastal evolution as demonstrated by the field evidence gathered after the event.

2.3 Erosional coasts

The input of energy, whether derived from solar, seismic or cosmic sources, into a coastal system via waves constitutes one of the main forces driving physical coastal processes. When considering coastal processes and resultant landforms, it is logical to begin with erosional processes and landforms, as it is erosion that is largely responsible for providing sediment for depositional processes, and in this way depositional coastal systems are to a significant extent dependent on mass input from erosional systems.

The ability and extent to which waves may physically erode a coastline is dependent on three sets of variables:

- *Wave environment of the coastline.* This includes orthogonal fetch direction, significant wave-height, frequency of high magnitude wave events, such as storms and tsunami.
- *Geology of the coastline.* This includes rock-type or **lithology** with its inherent hardness and susceptibility to physical and chemical weathering, geological structure

(bedding planes, joints, faults and folds), provision of rock suitable for use as erosional tools by waves, and inherited geological weakness, such as weathering associated with past climates and former sea-level situations.

- *Morphology of the coastline*. This includes areal configuration and inherited topography (headlands and bays promoting refraction), cliff heights and slope angles, and submarine topography (again influencing refraction).

In nature, however, it is not only waves that erode coastlines: tidal action, subaerial conditions (e.g. climate) and biological activity, including the activity of humans, may make a contribution, which will be discussed later.

The mechanisms of wave erosion are varied and have attracted the attention of many authors who have established a number of terms to express these processes. Waves breaking onto bedded, jointed or faulted rocks can create **hydraulic pressure** in these structural voids, which may lead to weakening and readying of rocks for detachment by the process of **quarrying**, where a wave removes loose blocks. Detached blocks may break down under wave activity through processes of **attrition**, such as **abrasion**, which involve the rubbing of rocks against one another, gradually rounding the rocks and reducing their size. Sediment provided by quarrying and attrition may then be used as erosional tools by waves to further erode rock through **corrasion**, which is the mechanical weathering of rock surfaces by abrasion.

2.3.1 Cliffs

Perhaps the most well known of all coastal erosional landforms are cliffs, which may loosely be defined as any steep coastal slope affected by marine processes, although subaerial processes also play a significant role in cliff formation (Hampton and Griggs, 2004). The form that a cliff takes may be determined by the following:

- *Inherited characteristics*. The sea may rework steep slopes initially formed by non-marine processes under different sea-level situations, for example some **plunging cliffs** that rise abruptly from deep water were originally the sides of now submerged glaciated valleys.
- *Wave activity*. Sufficiently energetic waves are required not only to erode rock, but also to remove debris created by wave erosion and material deposited at the foot of the cliff by subaerial processes operating higher up, which would, if not removed, obscure and protect the cliff, restricting its development.
- *Geology*. The hardness and structure of the rock is very important in cliff development, as some soft rock may fail to support steep slopes altogether, regardless of the level of wave-energy. Also, rocks with bedding planes that incline or dip seaward may slip off into the sea, with the resultant cliff angle determined largely by the dip angle, whilst rocks that possess horizontal bedding or dip landward may be capable of developing near-vertical cliff faces (Fig. 2.9).

Cliff retreat often proceeds through the formation of a **wave-cut notch** at the base of the cliff, which effectively undermines the cliff leading to slope failure, either in the form of rotational slumping or vertical cliff collapse. The notch itself is formed through quarrying and corrasion, and is best formed where waves actually break onto the cliff. Notches are unlikely to form on plunging cliffs, which simply reflect swell waves that it intercepts, or on cliffs fronted by a wide and low-angled shore which is capable of dissipating most wave-energy before it actually reaches the base of the cliff. Therefore, cliffs fronted by a

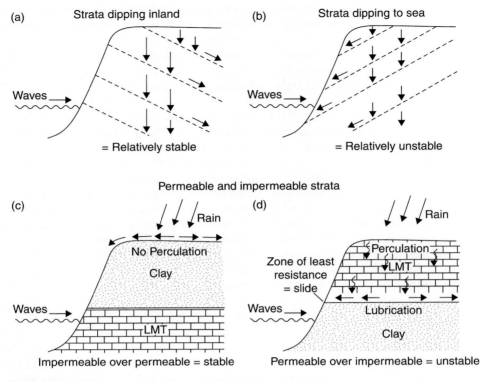

Figure 2.9 Influences of geological structure (a and b) and lithology (c and d) on coastal cliff development (LMT = limestone).

Source: French (1997).

narrow shore are likely to possess notches, and retreat in a cycle of notch formation, cliff failure and debris removal.

Other mechanisms of cliff retreat include the activity of landslides, mudflows and rotational slumps (Fig. 2.10a) in soft and weak lithologies, moving material from high up the cliff to the base in an attempt for the cliff slope to attain equilibrium. However, the subsequent removal of this material from the base of the cliff by wave activity destabilises the cliff and stimulates further mass movement, sustaining retreat. Also, cliff retreat can perpetuate itself through the development of **pressure-release jointing**. This occurs because rock expands as the confining pressure created by surrounding rock decreases as cliff retreat proceeds. Expansion takes place in the same direction as pressure release and so joints open up parallel to retreating cliff faces. Debris, water and organisms can invade these joints helping to prise them open, eventually leading to **toppling failure** (Fig. 2.10b), when large slabs of rock topple onto the shore.

Variations in rock strength along the length of a cliff can often lead to **differential erosion**, where weaker rocks are eroded at a faster rate compared to relatively stronger rocks. This may result in the formation of caves, arches and stacks. Caves are progressively excavated through the combined action of hydraulic pressure, quarrying and abrasion exploiting a weakness in the rock, such as a fault. A bridge-like arch is formed where a cave develops to such an extent that it emerges out on the other side of a headland. Whether during the cave or arch stage, the ceiling may become unstable through wave action and/or gravity, and collapse. If ceiling collapse occurs gradually, with vertical holes appearing as ceiling sections fall at different times, then **blowholes** may be formed, which may issue

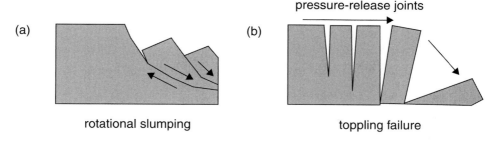

Figure 2.10 Mechanisms of cliff failure: (a) rotational slumping (arrows indicate relative movement); (b) toppling failure.

a cliff-top water geyser under favourable wave conditions. A stack is created upon the complete destruction of an arch's ceiling, so that it is no longer attached to the main cliff line and is essentially an island (Fig. 2.11).

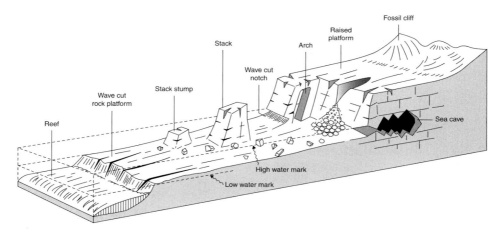

Figure 2.11 The geomorphology of a rocky coastline.

Source: Briggs *et al.* (1997).

Case Study Box 2.3

Beachy Head – cliff collapse in southern England

Beachy Head is a well-known chalk cliff along the Sussex coast near Eastbourne in southern England. On the night of Sunday 10 January 1999, a very large section of the chalk cliff collapsed. The actual dimensions of the cliff section that collapsed are uncertain; the Environment Agency (1999) suggest a 150 m high slab comprising hundreds of thousands of tonnes of chalk, but an item in the Geological Society of London's magazine (Nield, 1999) reports a more conservative estimate of a 60 m high and 15 m thick cliff section. More recent estimates suggest that 150,000 m³ of chalk debris was involved in this cliff collapse (see papers in Mortimore and Duperret, 2004).

The Environment Agency (1999) speculate that a series of climatic events may have contributed to the collapse:

- for 30 months prior to the autumn of 1998, southern England suffered a drought which may have weakened the chalk through drying;
- then through the autumn and winter of 1998 continued heavy rainfall is thought to have saturated the chalk;
- severe storms affected the coast of southern England for 12 days over the Christmas period, perhaps weakening and undermining the base of the cliff;
- cold conditions followed the storms with sub-zero temperatures, and freezing within the chalk may have produced further internal stress that provided the trigger for collapse.

It is highly probable that these climatic events contributed in some way to this cliff collapse, but these cliffs have been actively retreating since the rise of post-glacial sea level, so that the scale of this particular collapse is unlikely to be unique in the post-glacial erosion history of Beachy Head, and in my opinion does not reasonably constitute 'solid evidence that climate change, which was predicted as a result of global warming, had arrived with a vengence', as reported by the Environment Agency (1999).

2.3.2 Shore platforms

A consequence of cliff retreat is the creation of intertidal **shore platforms** that extend seaward from the base of the cliff (Stephenson, 2000). Although shore platforms are initially created by wave quarrying and abrasion activities, it is now appreciated that continued platform development is often aided by **bio-erosion** and weathering, particularly where tidal exposure is significant, and hence the abandonment of 'wave-cut platform' as a descriptor for these landforms. Also, shore platforms are seldom horizontal, and often possess a gentle seaward slope of up to $3°$, sometimes with small cliffs around the low-tide level and relatively steep ramps at the high-tide level. Rock or tide pools are also common on shore platforms, being contained within hollows excavated by quarrying and weathering.

The dimensions of shore platforms, under stable sea-level conditions, are intimately related to wave processes (Fig. 2.12a). Platform width, although associated with the rate of cliff retreat, has a finite limit (Trenhaile, 1999). This is because as a quasi-horizontal platform develops in front of a cliff, it increasingly acts to dissipate wave-energy before it reaches the base of the cliff, and thus beyond a critical platform width waves are unable to erode or remove debris protecting the cliff face, halting platform expansion. Under such circumstances, the only way in which cliff retreat can be maintained is if the height of the platform is reduced relative to sea level, so diminishing its dissipative effect. However, there is also a limit to platform lowering, because shear stress between the platform and passing waves decreases with water depth, so that at a critical depth wave activity becomes unable to erode the platform. Tidal range also has a part to play, with the potential to develop wider platforms under increasing tidal ranges (see section 3.3).

A number of classification schemes have been proposed for shore platforms. They are often classified according to their position within the tidal frame. This has given rise to

(a)

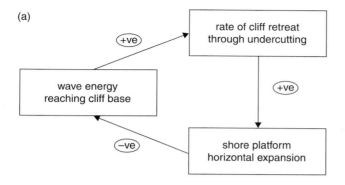

(b)

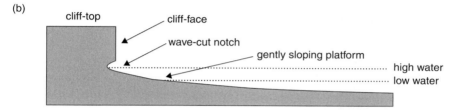

Type A (sloping) shore platform profile

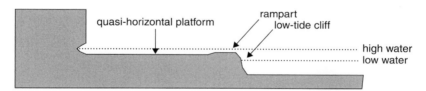

Type B (quasi-horizontal) shore platform profile

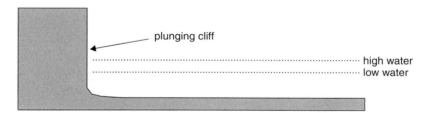

Type C or plunging cliff profile

Figure 2.12 (a) Negative feedback system between wave-energy, cliff retreat and platform expansion: high wave-energy leads to high rates of cliff retreat, but consequent platform expansion increases dissipation, reducing wave energy at the cliff and restricting further retreat. (b) Profiles of shore platform morphology types.

quasi-horizontal low-tide and high-tide platforms, or intertidal forms which slope from a high-tide level to low-tide, perhaps terminated by a low-tide cliff. Alternatively, shore platforms have been categorised by morphology, whether sloping, quasi-horizontal or near-vertical (types A-C of Sunamura, 1992) regardless of tidal position (Fig. 2.12b; Plate 2.3).

Plate 2.3 Types of shore platforms. (a) Type A or sloping shore platform and cliffs in well-bedded alternating Lias limestone and shales (Lower Jurassic) of the Glamorgan Heritage Coast (Wales, UK). (b) Type B or quasi-horizontal shore platform at Coledale Beach south of Sydney (New South Wales, Australia).

2.3.3   Ecology of rocky shores

Rocky shores represent quite an extreme environment, characterised by the following features:

- *Continual erosion of cliffs and platforms*. Provides dynamic and only very temporary habitats; for example, nesting sea-birds and salt-tolerant flowering plants colonising cliffs regularly have to re-establish themselves after cliff-falls.
- *Wave activity*. This makes attachment a problem that has led to the development of clinging strategies for many species of algae and invertebrates, although even these might fail in severe wave conditions. Wave activity also largely prevents the accumulation of nutrients, and most organisms here extract their nutrient requirements directly out of sea-water.
- *Highly variable environmental conditions*. This particularly concerns salinity, temperature and pH ranges, and water, oxygen and light availability. Tidal factors, especially drying/wetting (desiccation), freshwater seepage from cliffs reducing salinity, and water **turbidity** and depth at high tide limiting light are important.

The distribution of organisms on shore platforms is often zoned (Table 2.1). Because many intertidal species inhabitating these shores gain nutrients from the water, productivity is greatest near low tide. Here species richness, community complexity, and competition are high, but all decrease towards the high-tide level. Platform morphology influences this tide-related zonation, as a high platform slope angle would accentuate a zonation, whilst environmental conditions on a horizontal platform may remain spatially uniform at all states of the tide, negating zonal establishment.

2.4 Coral reef coasts

It is logical to consider coral coastlines here because the majority of reefs are composed of dead coral limestone, as only a thin layer of the reef surface supports living coral. This coral limestone is created by the activity of both small animals called polyps, which build delicate limestone structures that become the coral colony, and microscopic coralline algae that cement the delicate coral structures into a hard limestone pavement. Dead coral is susceptible to erosion by wave activity, which can reduce the limestone to rubble, termed skeletal sediments (Perry, 2007), that may be transported by wave-induced currents to infill active reef structures, or to create rubble mounds suitable for new coral colonisation, or to be swept up into a pile to create coral islands or **cays**. Therefore, wave activity plays an important role in coral reef erosion, deposition, and general morphological development.

Many of the world's coral reef systems have existed for millions of years; for example, parts of the Great Barrier Reef on the eastern seaboard of Australia have existed for 18 million years. However, their development has been influenced by the rise and fall of sea levels through **Quaternary** interglacial and glacial stages, respectively (Larcombe and Carter, 1998; see also section 5.1). During sea-level highstands coral building on the continental shelf has occurred, but during lowstands the continental shelf becomes terrestrialised and the reefs become limestone hills. Limestone is prone to dissolution, and **karst** features, such as caves, develop at these times. Under subsequent high sea levels, corals re-establish themselves, leading to cyclical reef evolution through time.

There are many forms of coral reef; however, there are three main types:

Table 2.1 The zonal distribution of characteristic organisms (lichens, algae, molluscs, and some other organisms) in relation to tide levels on a rocky shore (Cremona, 1988)

Tide levels	Zone	Environment	Lichens	Algae	Molluscs	Other organisms
	splash zone	high salinity, temperature extremes, desiccation, limited water	orange (*Xanthoria parietina* & *Caloplaca marina*), grey (*Ochrolechia parella* & *Lecanora atra*), and green lichens (*Ramalina siliquosa*)	*Prasiola stipitata*	*Littorina neritoides*	
high tide	upper shore	wide temperature variation, drying out at low tide	black lichen (*Verrucaria maura*)	*Pelvetia canaliculata, Fucus spiralis, Lichina pygmaea, Enteromorpha intestinalis,* and *Porphyra umbilicalis*	*Littorina saxatilis* to *L. rudis*	acorn barnacles
mean tide	middle shore	variable temperature, light limited at high tide		*Ascophyllum nodosum, Fucus vesiculosus, Cladophora rupestris, Ulva lactuca* and *Ceramium rubrum*	*Gibbula umbilicalis, Littorina littorea, L. littoralis, L. obtusata, Patella vulgata, Mytilus edulis,* and *Nucella lapillus*	*Actinia equina*
	lower shore	stable environment, high competition, light limited by water depth and turbidity at high tide, high wave activity		*Fucus serratus, Corallina officinalis, Lithophyllum* spp., *Himanthalia lorea,* and *Codium tomentosum*	*Patella vulgata* and *Nucella lapillus*	*Halichondria panicea, Anemonia viridis,* and *Bunodactis verrucosa*
low tide	sublittoral	stable environment, high wave activity, light limited		*Laminaria* spp., *Chondrus crispus, Gigartina stellata, Rhodymenia pseudopalmata,* and *Saccorhiza bulbosa*	*Gibbula cineraria, Calliostoma zizyphinum,* and *Patina pellucida*	

- **Ribbon reefs**. These are generally very narrow coral reefs that often occur on the seaward edge of reef areas. In the northern Great Barrier Reef, ribbon reefs develop on the edge of the continental shelf and may represent the upward growth of fringing reefs from a previous sea-level lowstand.
- **Fringing reefs**. These reefs develop along the coast of the mainland and of continental islands, and are amongst the only reefs accessible from the land without the use of a boat, and consequently are vulnerable to visitor pressure.
- **Patch reefs**. These are often oval to round reefs that occur on the continental shelf and develop through a series of six stages: (1) an ancient reef platform is re-activated by rising sea level to produce (2) a submerged reef that develops into (3) irregular reef patches around the platform periphery; these patches merge to create (4) crescentic reefs that eventually form a circle (5) enclosing a lagoon; sediment infilling of the lagoon leads to the formation of (6) a planar reef surface upon which a cay island may develop. The vast majority of coral reefs on the Great Barrier Reef are of this type.

The morphology of these reefs usually comprises an eroded fetch facing the 'windward' side and a relatively protected 'leeward' side characterised by the development of new coral colonies, often with massive stacks called 'bommies' which are composed of *Porites* coral (this does not apply to fringing reefs that have landward sides that adjoin the coast). Between the windward and leeward/landward sides, a reef platform develops that may often possess lagoons. Debris derived from erosion of the reef on the windward side is often deposited on the platform surface, graded from coarse to fine debris towards the leeward side. Wave refraction on subrounded patch reefs can transport this debris to the point on the reef platform where the wave orthogonals intersect, sometimes accumulating enough debris at these locations to begin the process of cay formation (Fig. 2.13).

2.4.1   Coral cays

Coral cays develop as a consequence of sediment build-up on a reef platform, perhaps initially as nothing more than a sand or shingle bank. However, as sediment continues to accumulate through wave and also tidal action, it may reach a point at which it is submerged during only one or two high tides a year. At this point, sea-birds may begin to nest on the cay bringing seeds with them, to complement those arriving by floating from mainland or continental island sources, and guano to help fertilise the cay sediment. Dune and mangrove vegetation may then develop, contributing organic matter to the developing cay soils and also trapping and allowing sediment to accumulate above high tide, leading to further cay expansion. When the cay attains a suitable size, the sandy cay substrate may

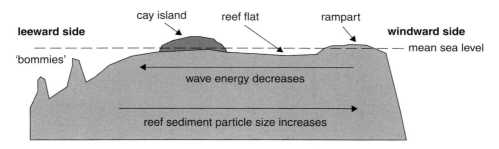

Figure 2.13 The geomorphology of a coral patch reef.

accommodate a freshwater reservoir that enables halophobic (salt-intolerant) plants, such as trees, to become established.

Vegetation helps to stabilise the cay, but this is also aided by the development of relatively hard and cemented cay sand. This may develop in two ways, either:

1. by the downward percolation of precipitation of phosphate from sea-bird guano to produce a hard pan called cay sandstone, which is sometimes of economic importance; or
2. by the precipitation of calcium carbonate in the pore spaces between the sand grains, so cementing them together to form **beachrock**.

These cemented sands underlie cays and serve to increase cay stability. However, at all stages of cay development, the cay is vulnerable to severe wave activity, such as storm waves associated with tropical cyclones. In many cases, cays have been destroyed by a single storm event, but quite often the cay is able to respond to short- and long-term wave variations by changing its morphology, eroding in one place and depositing in another. Therefore cays, like many coastal landforms, are considered to be **morphodynamic** structures, being able to change morphology in response to the dynamics of their environment. In this way, cays are often shifting, seeking to obtain equilibrium with the prevailing wave environment at a given time. In the field, morphodynamic changes may be detected by the presence of a number of criteria (Plate 2.4):

- the exposure of beachrock around cay margins indicates the loss and removal of overlying unconsolidated cay sand by wave erosion and transport;
- the presence of cliffs (usually ≤1m high) backing cay beaches indicates that wave erosion and removal of unconsolidated cay sand by wave transport is actively occurring;

Plate 2.4 Exposed beachrock and fallen trees, indicators of coral cay instability at Green Island, Great Barrier Reef, Australia.

- the occurrence of fallen trees with exposed roots on beaches indicates active under-mining of cay vegetation by wave activity (this is usually seen in conjunction with cliff formation).

In contrast to reef build-ups, which may be several million years old, coral cays are only a few thousand years old. This is because unconsolidated cay sand is not easily preserved between sea-level highstands. Also, within any given interglacial highstand, cays develop best after the stabilisation of sea level and the expansion of reef platforms. This is because under rising sea levels reef growth is primarily vertical, attempting to keep pace with sea level; however, horizontal growth is stimulated when sea level stabilises. Subsequently, reef platform expansion provides the material and foundation for cay development.

2.4.2   Coral reef ecology

Although polyps are animals, the availability of light is of crucial importance in the formation of reefs. This is due to the symbiotic relationship between polyps and tiny green algae, called zooxanthellae, that they play host to. Polyps feed by catching microscopic floating prey with their stinging tentacles; however, nutrients are supplemented by photosynthesising zooxanthellae, that also utilise some polyp waste products, such as carbon dioxide. Corals occur at all latitudes of the world, but it is only in relatively warm waters of $18°C$ and higher (ideally around $26°C$) that they are able to construct reefs. It is the combination of warm water, normal salinity, zooxanthellae and good light availability for their photosynthesis, that promotes high rates of coral production and reef development.

The symbiotic relationship between animals and plants on reefs is not restricted to coral polyps and zooxanthellae, as clams, sponges and sea squirts also have symbionts. Also, non-symbiont algae proliferate on reefs and are the food source for planktonic and benthonic invertebrates, and also for higher-order animals such as fish and turtles. These herbivores may then be preyed upon by carnivores. Photosynthesis in both symbiont and non-symbiont plants, therefore, forms the energy basis of the reef ecosystem. Bacteria consume organic wastes and dead plants and animals, which in turn are consumed by filter-feeders, such as sponges and clams.

Some animals actually prey on the coral, sometimes with serious consequences. The crown-of-thorns starfish (*Acanthaster planci*) is one such predator, growing up to 80 cm in diameter, with up to 21 arms covered with venomous spines. It is normally an uncommon animal inhabiting Indian and Pacific Ocean reefs; however, outbreaks occasionally occur when their populations explode to plague proportions, often numbering thousands to millions of individuals per reef. Under such circumstances an entire reef might be completely destroyed in two or three years. The reasons for such outbreaks are not fully understood and may be due to natural factors, human activity, or a combination of the two. The rate of reef recovery following an outbreak is variable: staghorn (*Acropora*) corals with a growth rate of 20–30 cm a year may recover within 20 years, but massive corals with a much slower maximum growth rate of 4 mm a year will take considerably longer to recover.

Coral reef ecosystems are extremely sensitive to environmental stress. A common sign of coral stress is 'bleaching' (Brown and Ogden, 1993; Brown, 1997). This apparent bleaching is brought about by the coral polyps ejecting zooxanthellae, and may be in response to:

- water temperature variations below and above coral tolerance levels – this may be a particular problem during pan-Pacific El Niño events that increase surface water temperatures in some regions;

- salinity variations brought about by increased freshwater run-off via rivers and streams into coastal areas of coral reefs; seaward reefs may also be affected where substantial rivers in flood jet freshwater out onto the continental shelf;
- increased turbidity due to high concentrations of suspended silt limits light availability for zooxanthellae photosynthesis – this problem is often linked to increased river run-off in association with changing land practices, such as deforestation and agriculture, which promote soil erosion; conversely, ultra-violet stress may also cause bleaching in non-turbid waters.

The increased frequency of bleaching events in recent years has led to speculation that global climate change may be partly responsible, and indeed the possible influence of climatic factors, such as El Niño and changes in some regional precipitation regimes, offers some support to this argument.

2.5 Barrier Islands

Barrier islands are depositional offshore linear features, separated from the mainland by a lagoon, and orientated parallel to the coast. They are usually comprised of sand that is built above the high-tide level and stabilised by vegetation. They are very variable in size, from tens to hundreds of metres wide, hundreds to thousands of metres long, and may support sand accumulations up to 100 m high. Typically, barrier islands occur on coasts with a low gradient and low tidal range, and thus over 10% of the world's coasts have developed barrier islands, including the Atlantic coast of the USA and some European coasts, such as The Netherlands. Adjacent barrier islands are separated from one another by tidal inlets. The inlets allow the exchange of water from lagoon to sea, and also facilitate sediment erosion, transport and deposition around the barrier island. Barrier islands are very important for defending vulnerable coastal lowlands behind them, principally by absorbing wave-energy and protecting the coastline from severe storm wave conditions. Therefore, there is strong interest in studying the formation and morphodynamics of barrier island systems.

Case Study Box 2.4

Impact of tourism on coral cay reefs – Green Island, Great Barrier Reef

Coral cays are often sites of tourist activity, which creates pressures that may affect cay morphology and influence its stability and vulnerability. Australia's Great Barrier Reef, a World Heritage Area, attracted around two million tourists in 2005 – a dramatic increase compared to around 750,000 visitors in 1993 (Great Barrier Reef Marine Park Authority, 2007). This increase in tourism, much of which is centred on cays, is largely due to the introduction of fast catamarans that are capable of rapidly ferrying tourists from the mainland, making day trips to the reef possible. Green Island, lying on a patch reef near Cairns, is one of the most popular cay resorts, attracting up to 319,000 visitors per year with peaks of around 1900 a day (Queensland Environmental Protection Agency, 2003). A survey of

Green Island since 1936 has shown a number of morphological and environmental changes.

- The position of a spit located at the western end of the cay has shifted from the northwest in the 1940s, to the southwest in the 1950s, and back again to the northwest in the 1970s, where it has remained. It is thought that this has little to do with tourist pressure, but is more concerned with cyclone activity, particularly cay streamlining in response to the prevailing cyclone direction.
- Also, on the western end of the cay, a groyne was constructed to protect the resort beach. However, this interrupted longshore drift patterns and reduced new sand input to the western beach, leading to localised erosion. In response to the erosion, the groyne was later dismantled, but then rebuilt following the artificial emplacement of 18,000 m³ of fine sand onto the beach by the Beach Protection Authority during 1974–76. It appears that by 1978, however, this sand had been lost from the beach to the relocated spit on the northwest corner of the cay.
- The area of the reef flat covered by algal mats and sea-grasses (e.g. *Halodule, Halophilia, Symodocea* and *Thalassia*) increased from 900 m² in 1945 to 130,000 m² in 1978, and has been interpreted as an effect of **eutrophication** from sewage released directly onto the reef flat. Indeed, algal and sea-grass colonisation was first apparent around sewage outlets. This is a problem not only for Green Island, but also many other cay resorts and fringing reefs close to mainland sewage outfalls. Measures to limit effluent pollution on Green Island have been in place since 1994 with the construction of a tertiary sewage treatment plant. Approximately 30% of the treated effluent, that was previously discharged into the sea, is now used to irrigate the grounds of the resort, but careful monitoring ensures that the island's natural aquifer is not contaminated. Also, water conservation is being promoted so that the amount of wastewater produced is reduced. For example, the installation of water-efficient toilets and showers in the tourist resort has reduced water consumption from 74 litres per person per day in 1995–96 to 55 litres in 1996–97, which in total is a change from approxmately 21 to 15 million litres.
- The total volume of sediment contained within the cay, which is dominated by foraminifera and the remains of other dead organisms (Yamano *et al.*, 2000), appears to have decreased and is thought to represent a negative feedback to generally increased eutrophy within the cay system. This is due to the sediment-trapping ability of sea-grasses that interrupts the normal sediment exchange between the reef flat and the cay. Sediment is often transferred from cay to reef flat during cyclones, but returned during fairweather wave conditions. Increased sea-grass colonisation on the reef flat traps this sediment and prevents it from being returned to the cay. Indeed, in 1998 approximately two-thirds of the Green Island periphery was characterised by exposed beachrock, back-beach cliffs, and fallen up-rooted trees indicating cay instability. In response, annual beach nourishment has been undertaken by the Queensland Parks and Wildlife Service since 1995 at a cost of around 10,000 Australian dollars per year.

2.5.1   Barrier island formation

The formation of barrier islands is controversial, with three principal hypotheses that may be applicable:

1. *Emerged-transgressive model.* Some barrier islands may represent offshore bars that were formed during the previous glacial sea-level lowstand; with the post-glacial rise in sea level they developed vertically using accumulated sediment transported onshore during the transgression.
2. *Submerged-transgressive model.* This refers to barrier islands that may have been coastal dunes during a lower sea level, but were isolated from landward coastal lowlands through submergence and transgression by post-glacial sea-level rise.
3. *Emerged-stillstand model.* The previous hypotheses consider barrier islands to represent the continued development of inherited features (offshore bars or dunes) from earlier sea-level lowstands; however, this theory suggests that barrier islands have developed since post-glacial sea-level rise stabilised, approximately 4000 years ago, to produce the current sea-level stillstand.

It is known that some barrier island deposits are certainly older than 4000 years, but it is possible that all three of these models may be correct in certain situations.

2.5.2   Barrier island morphodynamics

Barrier islands are highly dynamic environments and susceptible to changes in sediment supply, sea-level change, human interference, and wave energy, particularly linked to hurricane and storm conditions (Stone *et al.*, 2004). Mapping of barrier islands along the Gulf of Mexico (Louisiana and Mississippi) coast for more than a century has indicated the degree and types of morphodynamic changes that have occurred. McBride *et al.* (1995) have used this information to model the geomorphological response of barrier islands to natural and human variables (Fig. 2.14). The model identifies eight geomorphological response types:

1. *Lateral movement* involves the movement of sediment along the seaward front (seaside) of a barrier island, often characterised by erosion at one end and deposition at the other end of an island. This gives the impression that the island is moving laterally along the coast.
2. *Advance* refers to a shore that advances seaward through progradation in response to increased sediment supply or a lowering of sea level.
3. *Dynamic equilibrium* refers to a shoreline that appears to be stable over long periods of time, with neither significant erosion nor deposition.
4. *Retreat* applies to seaward-facing shores that retreat landward through the erosion and removal of sediment or from a rise in sea level.
5. *In-place narrowing* occurs where the seaward and landward (bayside) shores of a barrier island undergo erosion, leading to narrowing of the island, but with the core of the island remaining stationary.
6. *Landward rollover* often follows in-place narrowing, where the island beomes narrow enough for storm waves to overtop the island, eroding sediment from the seaward side to deposit it on the landward side. Due to this rollover of sediment an island will appear to migrate landward.

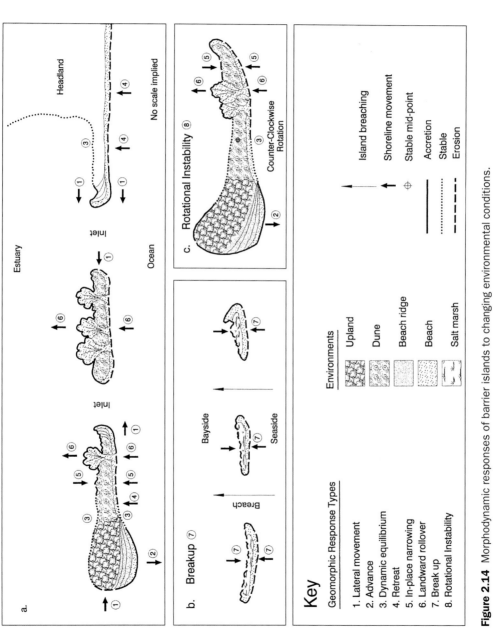

Figure 2.14 Morphodynamic responses of barrier islands to changing environmental conditions.

Source: reprinted with modifications from Marine Geology, **126**, McBride et al., Geomorphic response-type model for barrier coastlines: a regional perspective, pp. 143–159, 1995, with permission from Elsevier Science.

7. *Breakup* also often follows in-place narrowing, where narrowing has left the barrier island susceptible to breaching by waves to form new inlets, which may rapidly widen at the expense of barrier islands.

8. *Rotational instability* refers to an island that appears to be rotating, either in a clockwise or anti-(counter-)clockwise direction, in response to advance at one end and retreat at the other end of the island.

An example of barrier island geomorphological development according to the above response-type model is dramatically provided by the Isles Dernieres barrier system of the Louisiana coast in the southern USA (McBride *et al.*, 1991, 1995). It is situated on a sediment-starved coast that is experiencing a very rapid rise in sea level of 1 cm/year due to regional ground subsidence. For most of the 100 years up to 1989, the barrier system has experienced in-place narrowing, with 11.1 m/year of erosion and retreat on the seaward shore and 1.9 m/year of erosion on the landward shore. Breaches have subsequently occurred leading to the general breakup of the system. Eroded sediment is transported out of the system or stored subtidally as shoals in the ever-widening inlets. In this way, these barrier islands have been reduced in size by 78% over the study period, from 3532 to 771 ha, a rate of loss of 28.2 ha/year. Future projections based on the current rates show that this barrier system will disappear by the year 2020, although the impact of Hurricane Katrina in 2005 appears to have been low compared to elsewhere along the Gulf Coast. The disappearance of Isle Dernieres will have severe consequences for the Louisiana coastline that it previously protected, with estuaries and fragile coastal wetlands being exposed to the full intensity of hurricane conditions. This trend may only be offset through a programme of barrier island restoration through artificial sediment nourishment.

2.6 Beaches

Beaches are perhaps the most familiar and paradoxical of all coastal environments. They are composed of loose, unconsolidated, malleable material, sand and pebbles, and yet survive the roughest storms and wave conditions that affect coastlines. Yet it is exactly this characteristic that makes beaches so endurable: sand is highly mobile and can be moved about, and moulded into shapes that deal harmoniously with wave-energy, so ensuring their continued existence. Hard coastal structures, whether natural (such as cliffs) or man-made (such as harbour walls), are rigid and resist wave attack, and so are eroded – beaches are seldom eroded by waves, they simply metamorphose.

2.6.1   Beach profiles

The way in which beaches respond to wave-energy makes them excellent natural coastal defences, but the way in which they defend the coastline changes according to the level of wave-energy being received (Fig. 2.15). Under normal or **fairweather wave** conditions (low to moderate wave-energy) beaches have quite a steep gradient (e.g. 11° or 1:8 slope) and, as we discussed earlier, this steep gradient tends to reflect the moderate wave-energy back out to sea. The reason for this high beach gradient lies in the breaking of individual waves. When a wave breaks onto a beach, water travels up the beach as **swash** and after the swash has travelled as far up the beach as the energy will allow, the water returns to the sea under gravity, known as **backwash**. Under fairweather conditions, when there is a relatively long time between consecutive waves, the backwash will often return to the sea before the next wave breaks, and so does not interfere with the swash of the subsequent

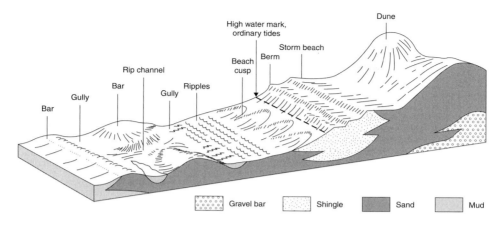

Figure 2.15 The general geomorphology and sedimentology of a beach.

Source: Briggs *et al.* (1997).

wave. Because swash energy is greater than backwash energy, and if the swash of the subsequent wave is not reduced by the backwash of the previous wave, there is a greater amount of sediment transported up the beach than down it. This builds up the beach, with waves taking sediment from low down on the beach and depositing it further up. It is not surprising then that fairweather waves are often called **constructive waves**. Fairweather conditions characterise the summer, and so steep beach profiles are often called **summer profiles** (also called **swell profiles**), frequently with a prominent ridge at the back of the beach called a **berm**, which marks the limit of the swash.

Under rougher storm conditions (moderate to high energy) beaches overall have a gentler gradient (e.g. 0.5° or 1:41 slope). This gentle incline tends to dissipate wave-energy because waves have a greater beach surface area over which to break. Often spilling breakers roll across these low-angled beaches for a considerable distance. Again the reason for the shape of this beach profile is due to swash/backwash interactions. This time waves arrive in relatively rapid succession at the beach, with the backwash of a previous wave returning down the beach whilst the swash of a subsequent wave is travelling up the beach. This reduces the ability of the swash to transport sediment, leading to net seaward sediment transport. In this way, sediment is taken from high up the beach and deposited seaward, thus reducing the gradient. Hence, these waves are often called **destructive waves**, and because they occur most commonly in the winter they produce a **winter profile** (also called **storm profile**). Low-gradient beaches often possess an extensive sand terrace around the low-tide level, which may have shore-parallel ridges or bars that separate similarly orientated depressions called **runnels** or gullies (Michel and Howa, 1999). Small rip channels breach the bars and connect up different gullies, allowing them to drain at low tide. Longshore sand bars may also occur slightly offshore.

Beach profiles, then, often change in response to seasonal conditions with complex inter-actions between weather, wave characteristics and beach geomorphology. Such seasonal changes may be seen on most beaches, but these are often superimposed on longer-term changes in morphology. For example, Fig. 2.16 shows beach profile changes on a gravel storm beach, which has been repeatedly surveyed using an electronic distance measurer over a ten-year period. Not only are there clear differences between profiles surveyed at different seasons within the same year, but also significant changes can be seen from year to year indicating the long-term evolution of the beach. It is clear that the beach is retreating

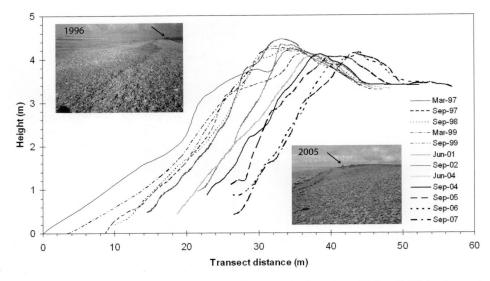

Figure 2.16 Beach profiles surveyed over a ten-year period (1997–2007) at Ru Vein, a gravel beach/barrier, in the Baie d'Audierne (Brittany, France). The profiles and inset photographs clearly indicate a narrowing and inland migration of the beach during this period. The arrows in the photographs point to the same building

Source: 2005 inset photograph kindly supplied by Dr Heather Winlow, Bath Spa University.

landward, but it also appears to be losing sediment volume and becoming smaller as a result. The beach faces the North Atlantic Ocean with a fetch in excess of 5000 km so it is likely that changes in wave environment will be an important factor to consider at this site. However, other factors, such as sediment availability, sea-level change and human activity, may also contribute to the changes observed. Each of these factors is discussed later, as is gravel beach dynamics, which help to explain the changes observed here.

2.6.2 Beach sediment

Beach sediment comes in all shapes and sizes, from rounded car-sized boulders to sand grains barely visible with the naked eye. Each sediment grain or particle, whether boulder or sand, can be described by its size, shape and origin. Words such as boulder, cobble, pebble, sand and silt are terms used to describe **particle size** (from largest to smallest). Shape can be described in terms of a particle's roundness and sphericity. These do not mean the same thing, as a rod-shaped particle (not at all spherical) may have well-rounded ends and sides, for example. The main shape categories are:

- *discs* (flat like a coin)
- *blades* (like a matchbox)
- *rods* (long and thin like a pencil) .
- *spheres* (like a ball).

A particle's origin can be described as:

- *lithogenic* (meaning *genic* = originating from *litho* = rock, and indeed particles of pebble size or larger are commonly simply lumps of eroded rock)

- *minerogenic* (mineral grains such as quartz, produced from the break-up of rock)
- *biogenic* (broken biological remains, such as seashells and corals).

Examining lots of particles together, as is usually the case with beach sediment, it is possible to describe the relationship of the particles to one another, otherwise known as sediment texture. Are the particles all the same size, i.e. is the sediment well-sorted by size? If the sediment consists of different-sized particles, such as pebbles and sand together, then it is poorly sorted by size. It is important when discussing **sediment sorting** to state what criteria are being referred to, as sorting according to shape also occurs, and sometimes sorting by both size and shape is evident. Also, what is the composition of the sand: is it purely minerogenic, or does it also have lithogenic and biogenic particles as well, and if so in what proportions?

The characters of sediment particles and texture reflect the wave environment of the beach and in turn can also influence the beach profile (Bascom, 1951). High wave-energy conditions are able to move nearly all particle sizes, and particle movement leads to contact with other particles, causing abrasion. Therefore, particles under these conditions are continuously being reduced in size, and yet somewhat ironically on high-energy beaches, such as Chesil Beach in southern England for example, there is very little fine sediment to be seen! To understand why this is so one must fully appreciate the dynamic relationship between waves and the beach, because the same wave-energy that erodes and creates fine sediment, at the same time prohibits its deposition, keeping the finer material in suspension in the water and transporting it to quieter waters where the energy levels are low enough for it to be laid down. Therefore, all beaches may be regarded as lag deposits, consisting only of sediment that wave-energy allows to be deposited, a bit like separating wheat from chaff, regardless of what is actually being produced on the beach by the same wave-energy.

The influence of sediment on the beach profile is significant, and is due mainly to the ability of the sediment to let water through it, that is its **permeability**. If sediment is highly permeable and allows water through it with ease, then under any wave conditions it may be able to allow the backwash of waves to return to the sea through the beach, rather than on the beach surface. This would eliminate the backwash and, as mentioned above, allow the swash of all waves to travel up the beach unimpeded, producing net sediment transport up the beach, building up the beach and increasing its gradient. The only way the beach gradient could be reduced in such circumstances is if the beach becomes waterlogged, perhaps by heavy rain or by very rapidly arriving waves. On the other hand, the backwash on beaches with low sediment permeability will return to the sea at or near the surface and may interfere with the swash, producing lower-gradient beaches.

2.6.3   Longshore beach features

As well as variations in the beach profile discussed above, beaches often vary alongshore, but unlike beach profiles, which are determined by on- and offshore wave currents, alongshore variation is influenced by alongshore currents, commonly referred to as **longshore drift** (Fig. 2.17a). Waves that approach a beach at an oblique angle may stimulate longshore drift through the oblique transport of sediment up the beach on the swash, but returning directly down the beach to the sea on the backwash under gravity, and in this zig-zag manner sediment may move alongshore until it is trapped and prevented from further transport. There are a number of natural sediment traps, of which the following are most important:

- **Re-entrant traps**. These are most often embayments bounded by headlands. Sediment introduced into such a bay may undergo longshore drift within the bay itself, but is unable to escape the bay due to the high wave-energy conditions affecting the headlands. Beaches that occur in such embayments are variably termed **bay-head beaches** or **pocket beaches**. If some sediment is allowed to escape around downdrift headlands, then a series of headland bound beaches may develop that expand in width downdrift as **fish-hook** or **zeta-form beaches**.

- **Salient traps**. As the name suggests, these traps project outward from the coast. **Spits** are examples that usually form where the coastline turns abruptly away from the longshore current pathway. Sediment continues to be transported and deposited linearly along the current pathway and does not follow the coastline. Spits are thus beaches that are anchored to, but largely detached from the shore, and act as a repository for drifted sediment. Also, bay-mouth bars may form where a spit spans a bay entrance from headland to headland (Fig. 2.17b).

- **Equilibrium traps**. The best examples of equilibrium traps are **cuspate forelands**. These are generally triangular features that may be formed by the convergence of two opposing longshore drift systems, with its point developing towards the minimum fetch direction. Local river sediment output and submarine topography may also be important in cuspate foreland development. This is certainly true for some related coastal landforms, such as **tombolos**, which are strips of sand linking offshore shoals and islands to the mainland, and may form by sediment drift and deposition in the low-energy conditions of the islands' shadow zone (Fig. 2.17c).

- **Deep sinks**. This collectively refers to sediment that is lost from the coast by transport into deep water, below wave-base, and so cannot be reintroduced to the coast except by sea-level fall. This is often aided by the presence of submarine canyons that intersect longshore drift paths and act as conduits for coastal sediment removal.

Straight coastlines are particularly affected by longshore drift, but as long as downdrift output of sediment is matched by updrift input, the beach system will remain in equilibrium. However, a reduction of updrift input, such as natural exhaustion or commercial dredging of the sediment source, will result in narrowing of the beach and increased vulnerability of the coast to erosion and marine inundation. Remedies for sediment retention include the building of **groynes** at right-angles to the shore to trap longshore moving sediment (Plate 2.5). These are often effective, but in turn reduce the amount of sediment reaching sections of the beach further downdrift.

Longshore features are also present on beaches where waves break parallel to the shore, although the amount of longshore transport is limited in this context. It might be difficult to comprehend at first why any longshore drift should occur on such coastlines; however, the onshore movement of water by wave translation needs to be balanced by an offshore flow. This is manifest in **rip currents** and ripheads, which are concentrated zones of offshore flow and are fed by the symmetrical longshore movement of water from both sides of the rip (Fig. 2.18). This establishes local water (and sediment) **circulatory cells** on the beach, where water movement is generally onshore between rips, whilst at the rips it is offshore, and between the two areas the flow is alongshore. Rip currents do also occur on beaches where waves approach obliquely, but tend to be asymmetrical, being fed from the updrift direction only.

Commonly circulatory cells are regularly spaced along a beach and are morphologically expressed on the beach as scallop-shaped depressions known as **beach cusps** (Fig. 2.15; Plate 2.6). A long-standing theory for the formation and rhythmical distribution of cusps lies in the superposition of incident waves and edge waves, created by reflection off a headland (see section 2.2). Where the crests of a shore-parallel incident wave and an edge wave intersect, wave-height will increase, whereas the intersection of an edge wave trough along the same incident wave crest will reduce wave-height, so producing an incident wave with variable wave-height along the length of its crest. The higher part of a breaking undulating wave crest will contain more water and have more power than the lower part of the wave crest, and so will travel further up the beach and carry more sediment with it. This extra deposition increases beach relief where the high breakers occur, which subsequently acts like a watershed channelling the backwash of the high breakers into the

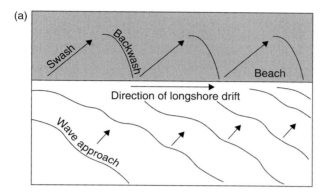

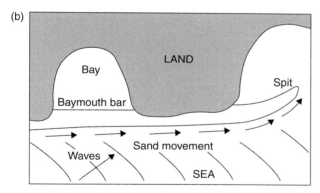

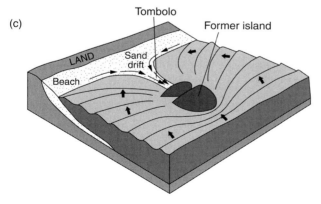

Figure 2.17 (a) Longshore drift. (b) Formation of spits and baymouth bars as a consequence of longshore drift. (c) Formation of a tombolo, an example of an equilibrium trap where two opposing longshore drift systems meet.

Source: Park (1997).

Plate 2.5 A groyne built to trap sand by interrupting longshore drift at Chapel St. Leonards along Lincolnshire's North Sea coast (UK). Notice the well-formed ripples which indicate strong current activity along this shore.

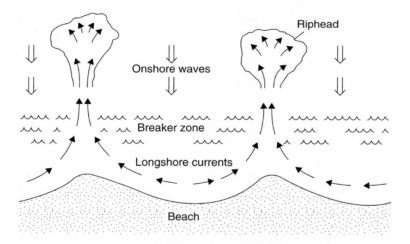

Figure 2.18 The development of rip currents along a swash-aligned beach.

Source: Briggs *et al.* (1997).

low breaker areas. Once this process is initiated it is somewhat self-perpetuating in that erosion and offshore transport of fine sediment occurs in the rips, whilst onshore transport of coarse material occurs at the built-up cusp horns. The horns also divide the swash, accentuating the circulatory cell systems. Below the water level, subaqueous topography is often a mirror image of the cusps, with rip-transported sediment building up deltas that occur opposite cusp embayments, and subaqueous hollows excavated by the high breakers providing sediment for the cusp horns.

Plate 2.6 Beach cusps formed in a gravel storm beach east of Nash Point, Glamorgan Heritage Coast (Wales, UK).

An alternative theory states that beach cusps are formed through positive feedback between existing beach morphology and swash characteristics, and is termed the self-organisation model (Werner and Fink, 1993). Irregular beach topography is enhanced by swash action so that, for example, a single upstanding positive feature on a beach would divide swash, locally creating concentrated backwash, rip currents and embayments on either side. Embayment erosion, in turn, implicitly creates a horn on the other side of the embayment, opposite to the original positive feature. It is argued that the internal dynamics of the beach-swash system (hence self-organisation) transforms these irregular occurrences into regular swash circulation patterns which lead to the development of rhythmical beach cusps. The alongshore spacing of beach cusps created in this way is considered to be related to the horizontal swash excursion, which is the distance travelled by swash up the beach, so that higher swash excursions result in more widely spaced cusps (Masselink, 1999). Indeed, more recent field evidence has shown that edge waves are not a prerequisite for beach cusp development (Masselink *et al.*, 2004).

Regardless of whether the edge wave or self-organisation model is responsible for beach cusp formation, the stability of beach cusps thereafter appears to be dependent on swash circulation. Masselink and Pattiaratchi (1998) explore the relationship between cusp prominence and spacing, swash excursion, and swash circulation (Fig. 2.19). They propose a model that relates three different fairweather swash circulation patterns to the further development of beach cusps:

1. *Oscillatory circulation.* This refers to swash and backwash that flow directly up and down the beach, respectively, and occurs when the cusps are too widely spaced or too subdued relative to the swash excursion. The outcome of this two-dimensional circulation is that sediment from subaqueous deltas in front of cusp embayments is transported onto the beach, gradually leading to sedimentary infilling of the embayments.

SWASH WATER CIRCULATION	DESCRIPTION
FAIR WEATHER CONDITIONS	
(a) 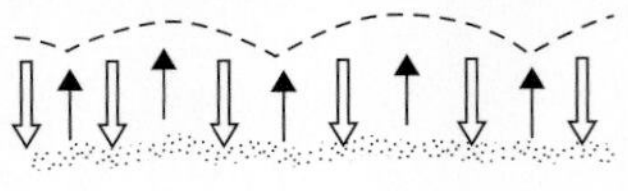	**OSCILLATORY** • Predominantly two-dimensional flow up and down the beach • Weak flow divergence on cusp horns • Weak flow convergence in cusp embayments
(b) 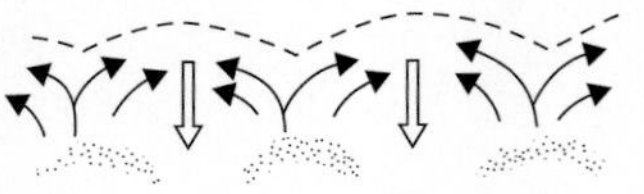	**HORN DIVERGENT** • Swash runup is diverted from cusp horn to embayment • In the embayment, flows meet to form a concentrated backwash • Mini rips form opposite cusp embayments
(c) 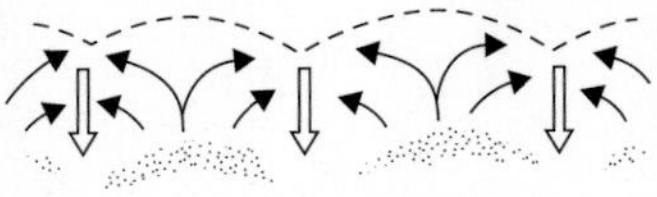	**HORN CONVERGENT** • Swash runup enters the cusp embayment with the bore front aligned with the embayment contours • Uprush spreads laterally to the horns and forms backwash • Mini rips may form opposite cusp horns
STORM CONDITIONS	
(d) 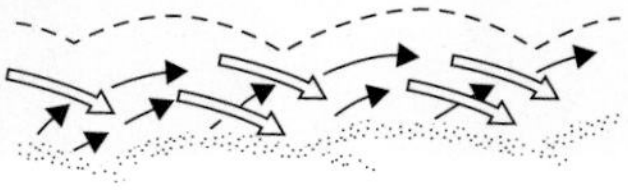	**SWEEPING** • Swash runup sweeps obliquely across the beachface • Backwash follows a parabolic arc • Littoral drift is pronounced
(e) 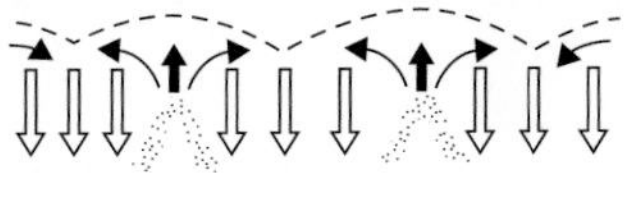	**SWASH JET** • In the embayment, strong backwash retards incoming swash until it has sufficient head to overwhelm the backwash flow and rush up the beach as a narrow jet • Swash runup in the form of a swash jet fans out laterally as in (c)

Figure 2.19 Beach cusp stability in relation to swash circulation.

Source: reprinted with modifications from *Marine Geology*, **146**, Masselink and Pattiaratchi, Morphological evolution of beach cusps and associated swash circulation patterns, pp. 93–113, 1998, with permission from Elsevier Science. Gerhard Masselink kindly supplied a copy of the original figure for reproduction here.

2. *Horn divergent circulation.* This type of circulation occurs when swash is in equilibrium with beach cusps, so that cusp horns divide the swash, concentrating backwash in embayments. This circulation pattern maintains existing beach cusp morphology.
3. *Horn convergent circulation.* This occurs when cusps are too closely spaced or too pronounced relative to the swash excursion, so that the swash essentially swamps the cusp features, with overtopping of the horns and wave uprush in the embayments which spreads laterally to converge at the horns. This rapidly promotes erosion of the horns and deposition in the embayments.

Also, two swash circulation patterns are given for storm wave conditions, sweeping and swash jet circulation, but both of these are destructive.

2.6.4 Gravel beaches

Gravel beaches (also known as storm or coarse-clastic beaches) are characterised by the dominance of large-size material (between 2 and 2000 mm diameter), and a steep shoreface, which usually means the shoreline is of a reflective type (Fig. 2.15). Unlike sandy beaches, the study of modern gravel beach dynamics has been neglected, mainly because it is difficult to deploy sensitive instruments in what are quite often high-energy and destructive environments (Carter and Orford, 1993). However, models of post-glacial gravel beach evolution are being refined with a growing appreciation of the fundamental controls involved in their development (e.g. Orford *et al.*, 2002).

Coarse-clastic shorelines are found mainly:

● in formerly glaciated (i.e. paralacial) or periglaciated areas (Plate 2.7a);
● on tectonic coasts where high-gradient streams deliver bedload to the shore (Plate 2.7b); or
● in wave-dominated areas subject to rock cliff erosion (Plate 2.7c).

They may be classed as:

● barriers that are free-standing or fringing, with a well-defined back-slope and back-barrier depression, enclosing lagoons and wetlands, and capable of migrating inland; or
● beaches that have no well-developed wave-formed landward-facing slope, and are often confined between headlands, as pocket beaches and abut cliffs, that can be further subdivided into **swash-aligned** or **drift-aligned** (i.e. perpendicular or oblique to wave approach, respectively).

The form of the shoreline is crucial to the morphodynamic status of gravel shorelines. There are two basic forms:

● the first comprises a single rectilinear slope from crest to wave-base (ignoring small ridges), which remains reflective under almost all wave conditions;
● the second comprises a concave-up form, which may exhibit a marked break in slope at or around the mid- to low-tide position, which alters the morphodynamic status of the shoreline as wave-height to depth and depth to wave-length ratios vary, as tide levels change. The break in slope is often mirrored by a change in sediment character.

A consequence of wave reflection is the development of edge waves, which on gravel shorelines are developed by:

● waves reflected off the shoreline and trapped by refraction (Carter, 1988); and
● waves resonating between headlands, usually associated with pocket and fringing beaches.

On the shoreline, the main manifestation of edge waves is the appearance of beach cusps, and the spacing of the cusps in this context is related to edge wave-lengths. Cusp morphology may exert a strong control over sediment movement and sorting. Furthermore, cusp development may dictate the pattern and position of barrier breaches and overwashing events during storms (Plate 2.8).

Permeability of the gravel shore is important in sediment transport. If the clasts are large enough, so producing sizeable interstitial pores, it may be that all swash sinks into the beach and returns to the sea through the beach. As discussed earlier, this will have the effect of minimising or eliminating backwash, so that net coarse sediment transport is landward. However, with large pore spaces, sediment decoupling may occur whereby fine sediment can be washed back through the beach to re-emerge and possibly to be deposited seaward, usually at a break of slope in concave-up beaches. Here, sand may be stored in the form of a sand terrace, and this may be added to by material introduced by longshore drift.

Plate 2.7 (a) Glacial till providing a rich source of coarse sediment for gravel beach development along the southern shores of Galway Bay (Ireland). (b) Gravel beaches along the Queensland coast north of Cairns (Australia) which are supplied with coarse sediment by high-gradient streams flowing off the uplifting coastal mountains. (c) Gravel barrier at Porlock (Somerset, UK) supplied by the updrift wave erosion of cliffs (also, notice the distinct berms formed on the beach face).

Plate 2.8 (a) A gravel barrier overwash fan at Ru Vein in the Baie d'Audierne (Brittany, France), and (b) a breach in the gravel barrier at Porlock (Somerset, UK) formed by severe storms on 28 October 1996.

Onshore sediment transport is concentrated within a narrow zone between breakers and the beach face and is dominated by **bedload transport**. Transport can involve either the movement of individual clasts or clast populations. Individual clasts on a flat shore are likely to move landward rapidly. Attached seaweed may aid this movement due to increasing clast buoyancy. Individual clasts tend to aggregate and, as accumulations of coarse particles develop, group-imposed transport controls are introduced. These controls influence sediment sorting, and there is a transition between an initial unsorted population of clasts to sorted subpopulations in terms of spatial distribution and size/shape characteristics. The net result is that gravel shorelines tend to become organised, in that they develop distinct cross-shore and alongshore **facies** that act to further limit transport. For example, more spherical clasts accumulate in the lower foreshore, whilst disc-shaped clasts tend to accumulate at or near the beach crest. Sorting such as this reflects wave-energy regimes, with most clasts being transported up the beach by the swash during high-energy storms, but then backwash preferentially transports the more mobile spherical clasts (spheres and rollers) back down the beach face (Williams and Caldwell, 1988).

2.6.5 The ecology of beach systems

Compared to many other coastal environments, sand and gravel beaches appear to be quite barren, and indeed they usually only support a living community of low abundance and diversity. Beach sediment, as we have seen, is highly mobile and regular disturbance occurs. Such an unstable substrate prohibits attachment by seaweeds that are so important in the food chain of rocky shores. However, plants do occur as microflora, such as bacteria and **diatoms**, attached to the surface of sediment grains. These are consumed by

---

Scientific Box 2.5

Human-induced destabilisation of gravel beaches

Gravel coasts, due to their morphodynamics, are extremely sensitive to human interference. Coastal developments, such as for tourist and recreational facilities, can leave gravel structures particularly vulnerable. For example, Jennings (2004) reports that recent harbour development at Eastbourne in Sussex (UK) has removed a number of gravel barrier ridges leaving only one seaward ridge. Such narrowing leaves the gravel system vulnerable to overwashing and breaching, and perhaps ultimately the general breakdown of the system. Orford *et al.* (1988) document a case of a gravel barrier at Carnsore in southern Ireland where such breakdown has occurred. Under natural conditions, fine sediment was supplied to the shore by an outlet stream from a back-barrier lagoon, which longshore drift distributed along the shoreline, creating a low-angled dissipative sand terrace in front of the gravel barrier. However, sand supply was artificially stopped by a dam being constructed across the lagoon outlet. Longshore currents eventually removed the seaward low-angle slope component and the disappearance of this dissipative element led to an increase in wave-energy reaching the reflective barrier. Cusp formation was initiated, which provided a template for overwashing, breach formation and general barrier degradation. Sediment was redistributed landward as washover fans.

Plate 2.9 Beach birds of southern California: during the winter months the quiet beaches around La Jolla (near San Diego) host many wintering wading birds such as (a) marbled godwit (*Limosa fedoa*) and (b) black-bellied plover (*Pluvialis squatarola*) as well as willet (*Catoptrophorus semipalmatus*), spotted sandpiper (*Actitis hypoleucos*), black turnstone (*Arenaria melanocephala*) and sanderling (*Calidris alba*) which all exploit the beach environment in slightly different ways.

microfauna or meiofauna living in the pore spaces between grains, which in turn support macrofauna. Because of the lack of shelter on beaches and the **infaunal** occurrence of microflora and fauna, most macrofauna, such as lugworms and shellfish, have adopted a burrowing lifestyle. These are largely preyed upon by wading birds (Plate 2.9) that have evolved specialised bills of varying lengths to probe into the sand, in addition to human exploitation of shellfish. Overall, the amount of organic matter residing in beach sediment is determined by wave activity, which may directly wash it away, and indirectly impair its storage by determining grain size, as larger grains have larger pore spaces which allow leaching and oxidization to occur. Therefore, low-energy beaches are often ecologically richer than high-energy beaches.

Problems within a beach do occur, such as the drying out of the upper beach at low tide, and the frequent disturbance by turbulent surf activity of the lower beach has led to ecological zonation, but it is not as clear and well-defined as in rocky shores. At and above the high-tide limit, plants rapidly colonise beach sediment, perhaps leading to dune formation on sandy beaches, stabilising gravel on coarser beaches, and generally increasing the abundance and diversity of species.

2.7 Coastal sand dune systems

Coastal sand dunes occur landward of the shoreline, usually above the high-tide level, and are often perceived to mark the landward limit of marine influence on the coast. The inclusion of sand dunes in a chapter on wave-dominated coasts at first sight may seem slightly inappropriate as dunes are mainly formed by aeolian (wind) processes and not by waves. However, many coastal dune systems are genetically related to sandy beaches that occur seaward of them, and these beaches are often characteristic of wave-dominated coasts as discussed earlier. They are extremely important coastal landforms as they often act as a coastal defence, protecting coastal lowlands from marine inundation. It is for this reason that coastal dunes are extensively studied from geomorphological, ecological and management perspectives.

A number of conditions are required for the formation of coastal sand dunes, including the following:

- An area landward of the beach that is able to accommodate blown sand, usually where cliffs are lacking.
- A strong onshore wind for transporting sand from its source on a beach to the dunes. For this reason, dunes are particularly common along coasts frequently affected by storm conditions, such as northwest Europe and northwest USA. Indeed, the height attained by a dune is determined by wind velocity, so that higher dunes occur in the stormiest regions.
- Suitably sized sand and an abundant supply of it. Without sand capable of transport by aeolian processes dunes cannot form. Also, for dune maintenance and continued development further supply of suitable sand into the system is needed. Hence, larger and well-developed dunes are commonly situated close to sediment sources, such as river outlets delivering sediment from erosive catchments to the coast.
- Vegetation to colonise and stabilise blown sand. Unvegetated dunes of the kind normally seen in arid deserts can occasionally occur at the coast, either where deserts are adjacent to the coast (e.g. Namibia, Africa) or where the rate of sand movement is high and prevents the establishment of vegetation (e.g. some dunes along the Washington and Oregon coast, USA). These unvegetated dunes are called **free dunes** and are sensitive to wind direction, often orientating themselves at right angles to the prevailing wind. However, most coastal dunes are colonised to varying degrees by vegetation and are called **impeded dunes**. Vegetation has a stabilising effect and to a large extent prevents sand loss and dune migration inland. The orientation of impeded dunes is more aligned with the source beach than with wind direction.
- A low gradient of the source beach coupled with a large tidal range. The combination of these two parameters means that large expanses of beach sand are exposed at low tide, increasing the time available for sand to dry and for the wind to transport it landward. Drying out of the beach is particularly important, and is one of the reasons why dunes are generally absent along tropical coasts, where high atmospheric humidity suppresses drying.

2.7.1   Aeolian sand transport and deposition

The initial movement of sand by wind occurs when a critical wind velocity is attained relative to a given sediment particle size; this is termed the **fluid threshold velocity** (Fig. 2.20b). The relationship is generally positive with increasingly high wind threshold velocities required to move coarsening particles. However, this relationship is reversed for very fine particles (clay and silt), which resist movement due to a high degree of cohesion between grains. Once initial sand movement has been established, transport may proceed as **surface creep**, where particles roll along the surface. Alternatively, if wind velocity is sufficient, sand may be entrained into the airflow and transported in suspension. A combination of these two processes produces the commonest mode of aeolian sand transport, that of **saltation**. This process involves particles that jump or hop then, after initially being stimulated at the fluid threshold velocity, are entrained into the airflow for a short distance before falling back to the surface under gravity. On landing, saltating particles impact with other grains and transfer kinetic energy, therefore reducing the

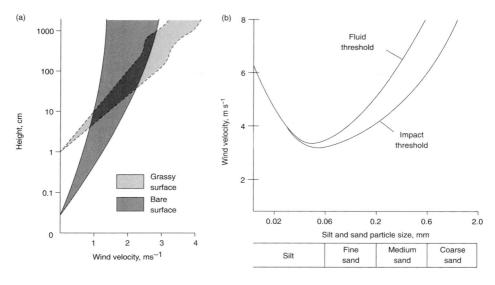

Figure 2.20 (a) Comparison of wind velocity over bare sand and grass (10 cm tall) surfaces, clearly indicating that vegetated surfaces prohibit deflation. (b) Fluid and impact threshold velocities for different particle sizes.

Source: Briggs *et al.* (1997).

threshold velocity required by the other particles to move. This lowered threshold is known as the **impact threshold velocity** and means that once sand transport is initiated it can be maintained by lower wind velocities (Fig. 2.20b).

The deposition of sand requires a reduction in wind velocity. On a beach this reduction occurs most commonly in the lee of obstacles, such as debris (e.g. seaweed, shells, pebbles, driftwood, litter) found along a strandline at the high-tide level. Sand accumulations quickly develop a streamlined duneform in response to the wind conditions, which is characterised by a gently sloping upwind stoss side and higher-gradient downwind lee side (Fig. 2.21). The internal sedimentary structure of dunes commonly exhibit co-occurring low- and high-angled bedding planes, known as **cross-bedding**, that represent old stoss and lee surfaces, respectively.

Dunes forming around debris build up to form low and unvegetated **shadow dunes**. Pioneer vegetation may subsequently colonise these mounds to form **embryo dunes** (Plate 2.10). The presence of vegetation traps further sand and allows the dune to develop substantially above and beyond the original piece of debris. In this way neighbouring embryo dunes can coalesce to form a dune ridge coincident with the initial strandline position, so forming **foredunes** running along the back of the beach.

2.7.2 Dune system morphodynamics

A single foredune ridge is the basic geomorphological requirement for the establshment of a coastal dune system. However, the morphodynamics of a dune system is largely dependent upon the nature of further sediment supply from the source beach (Fig. 2.22).

- If sediment supply is low then sand blown inland from the foredune may not be replaced, leading to a decrease in foredune volume and rendering it vulnerable to erosion, particularly cliffing or scarping of the foredune front by waves during storms,

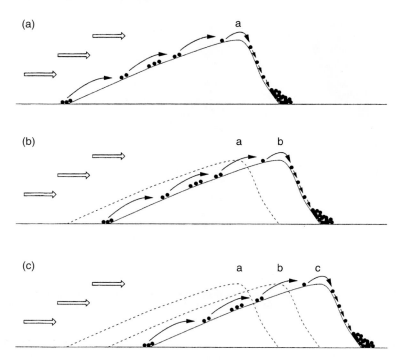

Figure 2.21 The formation (a) and migration (b and c) of sand dunes, also indicating the development of internal cross-bedding, where stoss and lee sides are preserved as oblique bedding planes.

Source: Park (1997).

Plate 2.10 Embryo dunes forming in discrete mounds at Barneville (Normandy, France).

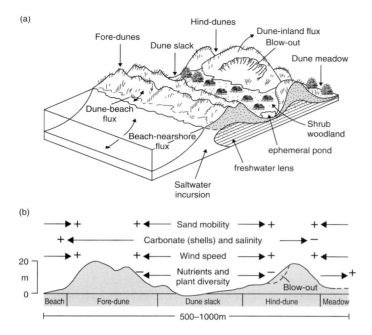

Figure 2.22 (a) Geomorphology of a coastal dune system. (b) The distribution and influence of some environmental parameters affecting coastal dunes (+ = increase; – = decrease).

Source: Briggs *et al.* (1997).

and fragmentation through the development of **blow-outs** along the foredune ridge. Blow-outs represent the localised removal of sand by the wind, a process called **deflation**, which creates relatively small hollows. Blow-out formation is often stimulated through disturbance of the dune by human activity (e.g. walking, riding), animal activity (e.g. rabbit burrows, grazing), or storm wave activity, which are all capable of undermining fragile dune vegetation, leading to sand liberation. Sand deflated from blow-outs is often redeposited downwind, where vegetation remains intact, in the form of crescentic **parabolic dunes**.

- Foredune morphology will be maintained if the loss of sand from the system is matched by new supply from the beach, so that there is no net change in sand storage within the dune.
- Where sediment supply from the beach is greater than the amount being lost from the dune (i.e. a net sediment gain), and where the coastal slope is relatively gentle, then dune systems may prograde seaward through the formation of new foredune ridges. In this way, a whole series of dune ridges may be formed, but only the most seaward is a foredune, whilst landward dune ridges are referred to as second, third and fourth hind-dune ridges, and so on. Depressions formed between the ridges are called **dune slacks**. These are often damp (**wet slacks**) where the surface of the slack intersects the water table, which generally limits deflation due to the presence of wet sand on the slack surface. However, water table fluctuations can sometimes lead to flooding in the winter and drying in the summer. The morphology of dune ridges and slacks is maintained by an undulating air flow or 'wind-waves' over the dune surface (Pethick, 1984). Continued progradation increasingly distances older dune ridges from the source beach so that eventually they may receive very little new sand. This leaves them vulnerable to blow-out formation which at its extreme may transform the initial

shore-parallel or **primary dune** ridge into a number of shore-normal **secondary dune** ridges that comprise sand reworked within the dune system.

2.7.3   Coastal dune ecology and management

Plants are of vital importance to the formation and stability of a coastal dune system (Fig. 2.20a). However, dunes provide a very harsh habitat, with a highly mobile sand substrate, very little water retention (except in wet slacks) coupled with the drying effect of the wind, high salt input, extreme ground temperatures, and minimal nutrient and organic matter content of the sand. Even so, at the high-tide level salt-tolerant pioneer plants colonise shadow dunes to form embryo dunes, and once these pioneers become established other plants are able to take hold, especially marram grass (*Ammophila arenaria*), leading to the ultimate development of foredunes. Foredunes are characterised by such marram grass or yellow dune communities, so-named because of the great amount of bare sand that is visible on the dune surface. This community is largely **xerophytic**, comprising plants that can survive conditions with minimal water, relying on their extensive rhizomous root system and thick shiny leaves to obtain and conserve water, respectively. Landward this gives way to grassland/heathland or grey dune communities, and eventually scrub/ woodland or climax communities. Wet slack communities are similar to freshwater bogs, marshes or wet meadows, and are often characterised by quite high floral and faunal diversity (Cremona, 1988).

Dune systems are extremely sensitive and disturbance that leads to destruction of vegetation more often than not results in deflation and blow-out formation. Blow-outs can seriously affect the integrity of the dune system and its coastal defence capability. Commonly, blow-outs in foredunes act as a template for breaching by storm waves, which may lead to flooding of lowlands landward of the dunes. Also, sand blown inland has in the past engulfed entire villages. Where sand supply is abundant, blow-outs are only temporary and the dune system is not necessarily at risk. Similarly, dune scarps commonly develop in the winter as storms erode the foredunes, removing sand to subtidal longshore bars, but then this sand is returned under the summer constructive regime. However, many dune systems have very limited new sand input, upsetting this state of dynamic equilibrium, and so are very susceptible to sand loss (Plate 2.11). Therefore, the management of coastal sand dunes is a major concern for many coastal protection agencies throughout the world.

One of the main sources of dune disturbance comes from their use as a recreational resource. A number of measures can be taken to minimise the effect of visitor pressure. Generally visitors to dunes are simply seeking access to a beach, and it is while they are making their way there that damage occurs. Therefore, only a few large car parks should be provided per dune system, rather than many small car parks, as this reduces the number of access points to the beach, so minimising potential foredune fragmentation. Walkways from these car parks should be managed to avoid pathway incision and the undermining of adjacent vegetation. This may be achieved through building boardwalks or other similar hardstanding structures, and also by prohibiting visitors from wandering off the walkways. Other activities, such as on-dune driving, horse riding and camping, should also be prohibited on vulnerable dunes.

Where disturbance has already occurred, some form of remedial action is required. Firstly, areas of bare sand within a blow-out, for instance, should be stabilised either by planting marram or by the use of artificial material, such as nets, collectively known as geotextiles. Secondly, sand deposition may need to be encouraged, mainly through the use of artificial sand traps, commonly in the form of picket fences that allow wind to pass

Plate 2.11 Dune scarp formed during winter storms and persisting into the summer, indicating limited new sand supply, at Genets in the Baie de Mont St Michel (Normandy, France).

through them, but reduce wind speed enough for deposition to occur. Such fences have the additional benefit of restricting access to these vulnerable areas.

Summary points

- Waves are created by solar, seismic or cosmic energy, and in deep water transfer only energy, and not matter.
- In shallow water waves transmit energy and matter, facilitating erosion of the coast, and transportation and deposition of sediment.
- Geomorphology of wave-dominated coastal systems is determined by the input and output of wave-energy and material (i.e. sediment). Coasts with high energy and low sediment input typically form erosional coastlines, whilst high sediment input systems tend to be depositional.
- The landforms and ecosystems of wave-dominated coasts are sensitive to interference from human activity.

Discussion questions

1. What is the geomorphological significance of nearshore wave modification processes?
2. Given that high wave-energy promotes erosion, why do high wave-energy beaches generally lack fine sediment?
3. To what extent are wave-dominated coastal systems influenced by human activity?

Further reading

See also

Geological and tectonic coastal classifications, section 1.3
The geomorphological significance of tidal range, section 3.4
Coastal responses to sea-level change, section 5.3
Approaches to coastal zone management, section 6.3

Introductory reading

Waves, Tides and Shallow Water Processes. Open University. 1989. Pergamon Press, Oxford, 187pp.
A thorough text focusing on the operation of coastal processes, including waves.

Tsunami: The Underrated Hazard (2nd edn). E. Bryant. 2008. Praxis Publishing Ltd, Chichester, 330pp.
An accessible text questioning the role of tsunami on coastal evolution and examining their geomorphological significance.

Coastal geomorphology. W. J. Stephenson and R. W. Brander. 2004. *Progress in Physical Geography*, **28**, 569–580; and W. J. Stephenson. 2006. *Progress in Physical Geography*, **30**, 122–132.
Two accessible reports of the same name that provide progress updates in fields relevant to this chapter.

World Atlas of Coral Reefs. M. Spalding, C. Ravilious and E. P. Green. 2001. University of California Press, 424 pp.
A colourful introduction to the biogeography of coral reefs around the world.

Islands at the edge. J. Ackerman. 1997. *National Geographic*, **192**(2), 2–31.
A colourful and informative account of the physical, ecological and human aspects of barrier islands.

Temperate coastal environments. A. Cooper. 2007. In: C. Perry and K. Taylor (eds) *Environmental Sedimentology*. Blackwell Publishing, Oxford, 263–301.
A good introduction to the sediments and depositional processes in temperate beach and barrier island systems.

Dynamics of beach-dune systems. D. J. Sherman and B. O. Bauer. 1993. *Progress in Physical Geography*, **17**, 413–447.
A useful review that considers beaches and dunes as interconnected coastal systems operating at different scales.

Advanced reading

Waves in Oceanic and Coastal Waters. L. H. Holthuijsen. 2007. Cambridge University Press, 404pp.
Relatively advanced text covering wind waves in detail.

Tsunami: wave of change. E. L. Geist, V.V. Titov and C.E. Synolakis. 2006. *Scientific American*, **294**(1), 56–63.
A useful overview of the Indian Ocean tsunami of 26 December 2004.

Rock coasts, with particular emphasis on shore platforms. A. S. Trenhaile. 2002. *Geomorphology*, **48**, 7–22.
A timely review of the processes and landforms of rocky coastlines.

Oceanographic Processes of Coral Reefs: Physical and Biological Links in the Great Barrier Reef. E. Wolanski (ed.). 2001. CRC Press, London and Boca Raton. 356pp.

A collection of papers on the physical, biological and management aspects of the Great Barrier Reef – useful case study material.

Mesoscale beach processes. D. P. Horn. 2002. *Progress in Physical Geography*, **26**, 271–289.
A thorough report on research progress in beach processes.

Concepts in gravel beach dynamics. D. Buscombe and G. Masselink. 2006. *Earth Science Reviews*, **79**, 33–52.
Presents a useful conceptual framework of dominant processes in gravel beach dynamics.

Coastal Dunes: Ecology and Conservation. M. L. Martínez and N. P. Psuty (eds). 2004. Springer, Berlin, 390pp.
A collection of chapters from a number of well-respected authors in this field covering a wide range of dune topics, from geomorphology to conservation.

3 Tidally-dominated coastal systems

For most coasts, the importance of tides lies in the way they help to determine the areal extent of the coastline that is affected by marine processes. Only in certain coastal environments, such as estuaries, do tides themselves and the currents they produce become a powerful agent of erosion and deposition. This chapter covers:

- **the generation of tides by the moon and the sun and the resulting tidal levels**
- **the importance of tidal range for coastlines and coastal processes, such as tidal current activity**
- **the hydrodynamics, geomorphology, sedimentology and ecology of estuarine systems, including associated coastal wetlands, such as salt marshes, mangroves and sabkas**
- **the impact of human activity in tidal coastal systems**

3.1 Introduction

Tides are a natural phenomenon that most people are aware of to a degree. They represent an everyday part of living at the coast and, indeed, many coastal activities (e.g. sailing, fishing, beachcombing) are strongly influenced by the state of the tide. But tides are cyclical and very predictable, and because of this people living at the coast must have accepted at a very early stage the control exerted by the tide on their lives and the pace at which they live.

All coasts are influenced to some extent by tides, but only a few types of coastal environment can be said to be tide dominated. Amongst those tide-dominated environments are estuaries, and these figure prominently in the settlement geography of the world. This is because estuaries are obvious trade routes and more often than not offer sheltered harbourage to sea-going trade vessels. Also, estuarine-fringing wetlands are usually flat surfaces which are ideal for urbanisation once they have been reclaimed and drained. Many major global urban centres are located in estuaries, such as London in the Thames estuary, so that although tide-dominated coastal environments may be regarded as uncommon, they support a disproportionate level of global human population. Thus estuaries are particularly affected by urban effluent discharge and other sources of pollution. This chapter introduces tides as they apply to all coasts, but then goes on to discuss in some detail the tide-dominated coastal systems of estuaries and their associated tidal wetlands, the salt marshes, mangroves and sabkas.

3.2 Tides and their generation

Tides are in fact waves with extremely long wavelengths. They result from the gravitational attraction of the sun and the moon, and although the sun is a much larger body, the moon is more influential because it is nearer the earth. As the sun and moon pass overhead they attract the surface ocean creating a **tidal bulge** (Fig. 3.1). Due to centrifugal forces, a second tidal bulge also occurs on the opposite side of the globe. When one of these 'tidal waves' meets the coast, the crest produces a high tide whilst the trough produces a low tide. The magnitude of tides produced in this way changes according to the relative positions of the earth, moon and sun during their orbital cycles. When the sun and moon are aligned with respect to the earth, when the moon is new or full, their component gravitational effects are combined to produce higher than average high tides known as **spring tides**. Conversely, when the sun and moon are at right-angles with respect to the earth, the gravitational attraction is dispersed, with the tidal bulge and the resulting high tides being lower than average. This is known as a **neap tide**. Similarly, the low-tide levels associated with spring and neap tides are respectively lower and higher than average. The spring-neap tidal cycle is approximately 14 days, so that it takes two weeks to go from spring to neap to spring tide. The largest of the spring tides occur at the vernal (spring) and autumnal equinoxes, when the sun crosses the equator.

Along most coasts, including nearly all open Atlantic coastlines, there are two high tides and two low tides every day, the **semi-diurnal tides**. High and low water are separated here by approximately 6 hours and 13 minutes, so that high and low water occur slightly later each day. However, some coasts experience only one high and one low tide each day due to local factors; these are **diurnal tides**, which occur, for example, at Do-Son in Vietnam and New Orleans in the USA. And along a few coasts the two are mixed, where the daily tides are dominated by one extreme high and low tide, but smaller secondary high and low tides do occur. This **mixed tide** is typified by tides along the western seaboard of the USA, such as at San Francisco and Los Angeles (Figs 3.2 and 3.3).

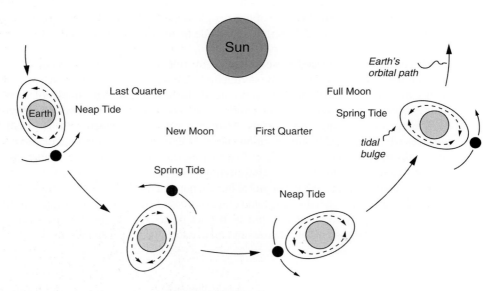

Figure 3.1 The formation of the tidal bulge with regard to the relative position of the earth, moon and sun.

Source: Briggs *et al.* (1997: 60, fig. 4.11).

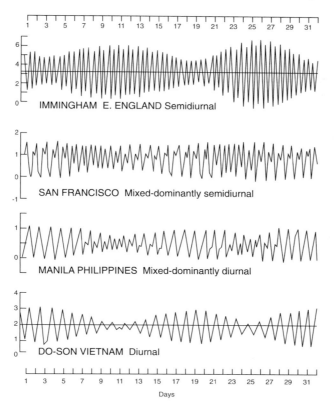

Figure 3.2
Examples of semi-diurnal, mixed and diurnal tidal cycles.

Source: Briggs *et al.* (1997: 60, fig. 4.12).

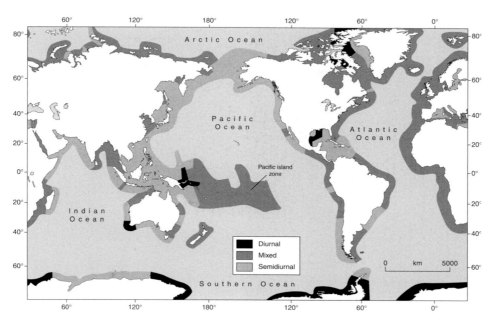

Figure 3.3 The distribution of semi-diurnal, mixed and diurnal tidal cycles around the global coastline.

Source: Briggs *et al.* (1997: 318, fig. 17.16).

It is clear from the above discussion that the regularity of orbital cycles produces predictable variations in tides at a number of different timescales from daily to decadel, and hence **tidal cycles**. Indeed, an important tidal cycle that can affect the establishment of tidal levels (e.g. mean sea level) takes 18.6 years to come full circle.

The spinning of the earth beneath the tidal bulge makes the bulge appear to travel uninterrupted around the globe, and on a landless earth it would do this. However, in reality the continents get in the way of the tidal bulge and, as with all intercepted waves, reflection occurs. In this way tidal bulges are pulled by the moon and sun westward across oceans, and upon meeting the eastern side of continents travel eastward back across the oceans as reflected waves just in time to meet the moon and sun coming round again. This simple to and fro scenario doesn't actually occur, however, due to the **Coriolis effect**, which under the influence of the earth's rotation causes travelling water, winds and other moving objects to appear to curve to the right (clockwise) in the northern hemisphere and to the left (anticlockwise) in the southern hemisphere – just look at water going down the plughole, it swirls clockwise and anticlockwise in the northern and southern hemispheres, respectively. This means that any given tidal wave travels in a circular manner known as **amphidromic motion**, and every ocean basin possesses an **amphidromic system**. An **amphidromic point** marks the centre point of an amphidromic system, and the positioning of these points in an ocean basin depends on the geometry of the ocean basin, including

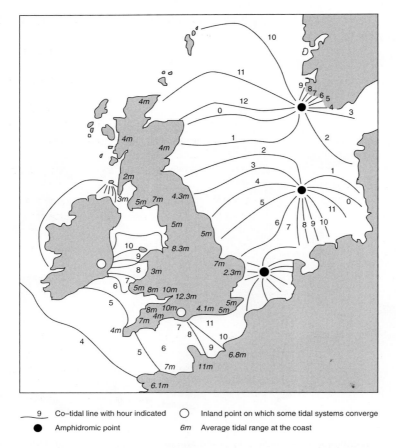

9 Co–tidal line with hour indicated	○ Inland point on which some tidal systems converge
● Amphidromic point	6m Average tidal range at the coast

Figure 3.4 The amphidromic systems in the seas around the British Isles.

Source: Briggs *et al.* (1997: 61, fig. 4.13).

coastal configuration and bathymetry, and they can even occur inland as degenerate or degraded amphidromic points.

Figure 3.4 shows, for example, the amphidromic systems around the British Isles. The radiating lines are co-tidal lines which show the position of the tidal wave in hours from the beginning of a tidal cycle. Therefore, it is clear that high water travels around the amphidromic point, so that low water occurs on the opposite side of the system to high water at any given time. This means that the water level at the amphidromic point does not change, remaining unaffected by the tidal wave. The further one travels away from the amphidromic point along a co-tidal line, the more extreme will be the tidal excursion from mean sea level, with increasingly higher high tides and lower low tides.

3.2.1   Tide and datum levels

Tidal cycles can be predicted and used to produce tide tables for particular locations, although these predictions are supplemented to a high degree by tidal measurements made by tide gauges. Tide gauges allow tidal levels to be established, some of which are employed as survey datums for making maps and charts. The most important of these levels are:

- *high and low water (HW and LW)* – the maximum and minimum tide attained during any one tidal cycle, respectively;
- *mean high and low water springs (MHWS and MLWS)* – the average spring high and low water levels, respectively, over a period of time;
- *mean high and low water neaps (MHWN and MLWN)* – the average neap high and low water levels, respectively, over a period of time;
- *mean high and low water (MHW and MLW)* – the average of all high and low water levels, respectively, over a period of time;
- *mean higher and lower high water (MHHW and MLHW)* – the average of the higher and lower high water levels, respectively, that occur in each pair of high waters in a tidal day (approximately 24 hours and 50 minutes) over a period of time;
- *mean higher and lower low water (MHLW and MLLW)* – the average of the higher and lower low water levels, respectively, that occur in each pair of low waters in a tidal day over a period of time;
- *mean sea level (MSL)* – average of water levels observed each hour over a period of at least a year, but preferably over a period of 19 years, so as to encompass the 18.6 year tidal cycle;
- *mean tide level (MTL)* – average of all high and low water levels recorded each day; this will usually differ only slightly from mean sea level;
- *highest and lowest astronomical tide (HAT and LAT)* – the highest and lowest water level, respectively, that is predicted to occur under any combination of astronomical conditions.

For land maps, either mean sea level or mean tide level are most frequently used as datums, depending on the detail of the tide gauge record available. In the United Kingdom, for example, mean sea level determined from six years (1915–21) of continuous tidal records at Newlyn in Cornwall is used by the Ordnance Survey as their datum, and called Ordnance Datum (Newlyn), although present UK mean sea level is approximately 0.1 m higher than recorded in the period 1915–21. Sea charts, however, use local low water or lowest astronomical tide as Chart Datum, so that below this level there should always be some water under the keel. Tide tables give tidal heights relative to Chart Datum so that mariners add predicted tidal heights to Chart Datum to obtain water depth at a given time. Because Chart Datum is based on local low water levels it does not represent a horizontal plane, so that the height difference between Chart Datum and mean sea level varies from place to place.

3.2.2   Meteorological effects and storm surges

Observed tidal levels may be significantly different from predicted levels. When this occurs it is usually attributable to meteorological effects. The ocean surface essentially acts like a barometer, with a rise or fall of the sea surface of 1 cm for every millibar (mb) of change in atmospheric pressure. For example, a high pressure atmospheric system 50 mb above average would lower sea level by 50 cm, whereas low pressure of a similar magnitude would allow sea level to rise by 50 cm. When low atmospheric pressure and predicted high tides co-occur, the tide level will be higher than anticipated and quite often the flooding of coastal lowlands takes place. This can happen on a very damaging scale when the low pressure is combined with storm conditions, with very high onshore winds. Where these storm winds blow water into semi-enclosed seas at the time of high tide, the water level can be piled up to several metres higher than predicted, resulting in a **storm surge**. Extensive coastal flooding can occur in association with storm surges, and are most devastating along low-lying coasts where their effects can extend many kilometres inland. Storm surges can be geomorphologically significant in that they may overtop or breach relatively low coastal features, such as dunes, and they often lead to wave attack at higher levels than attained by normal wave conditions.

Cyclones commonly produce storm surges along tropical coastlines (Box 3.1), but elsewhere they are relatively rare, although the North Sea due to its configuration is regularly threatened, and suffered a major event in 1953 that caused the loss of thousands of lives in eastern England and The Netherlands. This event stimulated the setting up of a storm warning service for eastern England and the construction of the Thames Barrier, to protect London from a similar disaster in the future. A similar situation exists along the coasts of the Gulf of Mexico where hurricane landfalls are often associated with storm surges.

Case Study Box 3.1

Storm surge hazards in the Bay of Bengal

The Meteorological Office in the UK report that in November 2007 a deep cyclone, named Sidr and labelled 06B, developed in the Bay of Bengal and tracked towards Bangladesh on the coast of the Ganges–Brahmaputra Delta. It made landfall in the Sundarbans mangrove forests of southern Bangladesh on 15 November, where winds peaked at 215 km per hour. Three million people are thought to have been directly affected, with the death toll as of 25 November standing at 3500, but was expected to rise. These deaths were not due, however, to the high winds, but to flooding from heavy rain and a severe storm surge. Some reports suggest that the storm surge was as much as 5 m high and severely damaged the Sundarbans World Heritage site, with experts estimating that it may take 40 years for the mangrove system to recover.

As devastating as Cyclone Sidr has been, the area has a long history of cyclone activity and storm surge flooding. The Indian Ocean accounts for 10% of the world's cyclones, with an average occurrence of 1.77 cyclones per year. These cyclones often track northwards into the Bay of Bengal, where the funnel shape of the bay piles up the water at the delta coast, and in conjunction with storm surges and increased precipitation, wind velocities and wave-heights, results in extensive flooding of the delta area (Fig. 3.5).

Historically, between 1797 and 1991 the area was hit by 60 severe cyclones, causing nearly one million deaths. The cyclone of 29 April 1991 was well documented and nearly 140,000 people died, most of whom were children under the age of 10, women and the elderly, and most drowned. Cyclone shelters had been constructed since the early 1970s but these could accommodate only 450,000 of the five million people at risk (Park, 1997). In 1991, one million homes are estimated to have been destroyed and another million damaged. Agriculture was severely affected, with 60% of cattle, 80% of poultry, 113,300 ha of standing crops, and 72,000 ha of rice paddies destroyed. Coastal flooding was enhanced because some 470 km of protective embankments failed; more recent protection measures are thought to have lessened the impact of Cyclone Sidr, but are still considered inadequate. Disease and famine affected the survivors, and the total economic impact of the 1991 event was 2.4–4.0 billion US dollars (Kay and Alder, 1999), whereas damage caused by Cyclone Sidr in 2007 has been initially estimated to have cost around 4.5 billion US dollars.

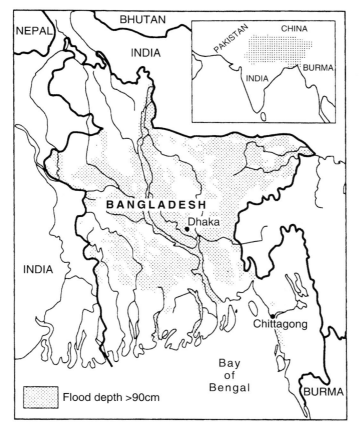

Figure 3.5
Areas of the Ganges–Brahmaputra Delta prone to flooding to depths greater than 90 cm.

Source: Pickering and Owen (1997: 321, fig. 8.7).

Bangladesh is not the only country to suffer storm surge impacts in the Bay of Bengal, as the Bay spans the international boundaries of India, Myanmar (Burma), as well as Bangladesh. For example, a 'super cyclone' made landfall in the Indian state of Orissa on 29 October 1999 with winds peaking at 255 km per hour and a storm surge apparently 6 m high that penetrated 14.5 km inland. Government statistics suggest ten million people died, but unofficial figures indicate a greater death toll. The local authorities have been criticised for a lack of preparedness regarding cyclone hazards in India prior to the event; however, response to a smaller cyclone in 2002 demonstrated that community hazard awareness had improved and new procedures put in place after 1999 were deployed to good effect (Thomalla and Schmuck, 2004).

3.3 Tidal range

The height difference between high water and low water during the tidal cycle is known as the **tidal range**. As discussed earlier, tidal range increases with distance from an amphidromic point, so that a coastline located near an amphidromic point experiences a small tidal range, whilst a coast on the periphery of an amphidromic system will experience a much greater tidal range. In addition, a number of other factors contribute to the great variety of tidal ranges experienced on the world's coasts. These include the following:

- *Bathymetry* – because of the enormous wavelength of the tidal wave it can be considered everywhere as a shallow-water wave. Therefore, it can undergo refraction like all waves and become focused on particular stretches of coastline, where tidal energy, height and range are increased.
- *Width of continental shelf* – the very shallow waters encountered by a tidal wave on a continental shelf reduce celerity and increase the wave-height. The slowing of the front of an approaching tidal wave allows the back of the wave to catch up, so increasing wave-height further. Therefore, wider continental shelves allow more time for the broad tidal wave crest to concentrate into a narrower but higher wave, increasing tidal height at the coast.
- *Coastal configuration* – tidal waves entering coastlines which are restricted in some way, such as embayments, gulfs and estuaries, will become compressed as they progress, increasing tidal height and range. This is most marked in funnel-shaped estuaries where the estuary width decreases incrementally upstream, such as the Severn Estuary in southwest Britain where the tidal range is in excess of 14 m. Conversely, open ocean coasts, often coupled with a narrow continental shelf, tend to reflect the tidal wave, resulting in minimal tidal ranges. Nichols and Biggs (1985) examined the influence of configuration on tidal range in estuaries. They state that variations in tidal range along an estuary are determined by the relationship between the upstream convergence of the estuary sides and the friction created between the tidal waters and the estuary bed (roughly equivalent to the surface area of the estuary), because increased friction will diminish the tidal range. For estuaries where the effect of convergence is greater than friction, tidal range will increase up-estuary, producing a **hypersynchronous estuary**, and in these estuaries a **tidal bore** wave may be produced, heralding the incoming tide. Where convergence equals friction then tidal range will be uniform along the length of the so-called **synchronous estuary**; and for

estuaries where the effect of convergence is less than friction, then tidal range will decrease up the estuary to produce a **hyposynchronous estuary** (Dyer, 1997).

Coasts can be classified according to their tidal range (Davies, 1964), with three categories being recognised (Fig. 3.6):

1. **Microtidal** coasts are those that experience tidal ranges of less than 2 m and are characteristic of open ocean coasts, such as the eastern seaboard of Australia and the majority of the Atlantic African coastline.
2. **Mesotidal** coasts possess tidal ranges between 2 and 4 m according to Davies (1964). Some authors, however, consider mesotidal coasts to have tidal ranges between 2 and 6 m (e.g. Briggs *et al.*, 1997), although the majority of texts adopt Davies' definition (e.g. Pethick, 1984; Carter, 1988; Summerfield, 1991; Viles and Spencer, 1995; French, 1997). Examples of mesotidal coasts include most Malaysian and Indonesian coasts, and the eastern seaboard of Africa.
3. **Macrotidal** coasts experience tidal ranges in excess of 4 m according to Davies (1964) and most other authors, but are defined as greater than 6 m by Briggs *et al.* (1997), a range considered by some authors to indicate hypertidal conditions. Macrotidal coasts occur where the continental shelf is wide, allowing the shoaling tidal wave to increase in height, and where the coastal configuration amplifies the tidal height. Examples include most of the northwest European coastal seas (e.g. Celtic Sea (including the Severn Estuary), North Sea, English Channel and Bay of Biscay), and parts of northeastern North America (e.g. Hudson Bay and the Bay of Fundy).

Macrotidal coasts are considered to be tide-dominated, in that most erosional, transport and depositional processes operating here are driven by tidal forces. Mesotidal coasts are considered mixed, with wave and tide processes being equally important. Microtidal coasts, however, are wave-dominated, and many of the coastal systems discussed in Chapter 2 are of this type. Overall, there is a close relationship between tidal range and the type of coastal landforms encountered along any given coastline, and therefore an appreciation of tidal range is essential in understanding coastal system diversity. For example, estuaries with their component tidal flats (whether sand or mud flats), salt marshes or mangroves and characteristic ecosystems are typical of coasts with high tidal ranges, whereas at the other end of the spectrum barrier islands (as discussed in section 2.5) are commonest along microtidal coasts.

   Tidal range is particularly important for coastal geomorphology because it influences the operation of physical processes (Fig. 3.7). There are a number of reasons why such importance is attached to tidal range:

● Tidal range and the gradient of a coast together determine the horizontal extent of the **intertidal zone**, which is the area lying between high and low water. High-gradient microtidal coasts have the smallest intertidal zones, whilst low-gradient macrotidal coasts have extremely extensive intertidal areas. Ecological diversity in intertidal zones is often greater and more complex on macrotidal coasts, with salt marshes occurring high and unvegetated tidal flats low within the intertidal zone.
● Tidal range determines the extent of the vertical distance over which coastal processes operate, especially wave activity. On microtidal coasts, wave breaking is concentrated within a very narrow vertical zone throughout the tidal cycle, and it is under these conditions that well-defined erosional features such as wave-cut notches are preferentially formed. However, wave-energy on macrotidal coasts can be distributed over many metres throughout the tidal cycle, so that its erosional capacity is relatively

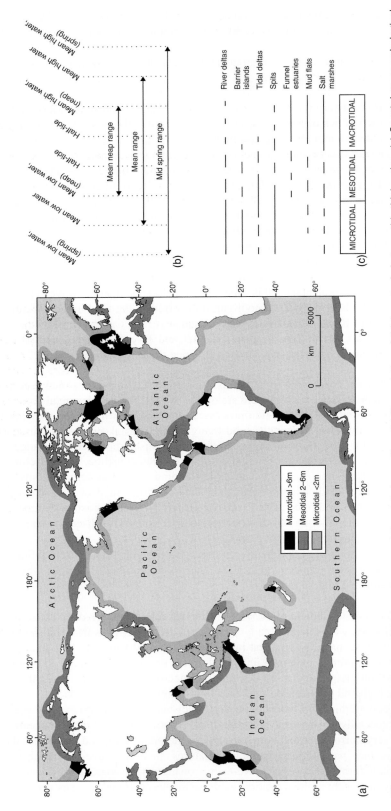

Figure 3.6 (a) The distribution of tidal ranges around the global coastline. (b) The variation of tidal range during monthly tidal cycles. (c) Coastal geomorphological features associated with the various tidal range categories.

Source: Briggs *et al.* (1997: 318, fig. 17.17).

diminished, but it does mean that wave activity influences a wider area. The same is also true for the operation of tidal currents (see section 3.4), with a greater tidal range subjecting a wider area to their activity.

● The periodic rise and fall of the tides causes wetting and drying of the substrate within the intertidal zone. Generally, the greater the tidal range the more substrate is exposed or submerged at different states of the tide. This is important for a number of processes, including salt weathering, where seawater invading hard crystalline or laminated rocks submerged at high tide is evaporated when exposed at low tide. Salt crystals grow within minute voids in the rocks, producing stresses that weaken and ultimately lead to the disintegration of the rock. This is particularly prevalent along tropical coasts, where evaporation and salt crystal growth can follow rapidly after exposure, and where especially susceptible rock types, such as granite, are found at the coast. Also, as discussed in section 2.7, sand dunes are more likely to develop on coasts with a relatively high tidal range, as this provides a wider expanse of beach sand for drying and subsequent landward tranpsort by aeolian processes.

3.4 Tidal currents

As water rises and falls with the tides it produces **tidal currents**. A rising tide that floods the intertidal zone is known as a **flood tide**, whilst a falling tide is the **ebb tide**. The significance of tidal currents lies in their ability to entrain and transport sediment. Maximum tidal current velocity is achieved at the flood and ebb tide mid-points, then at high and low water current velocity decreases to zero (slackwater) before reversing. Therefore, maximum sediment transport occurs at the mid-points in the flood and ebb tides (i.e. midway between high and low water and vice versa), whilst sediment deposition predominantly occurs around slackwater (i.e. at high and low water) (Fig. 3.8a). Critical tidal current velocity thresholds exist for transporting different particle sizes. Transport

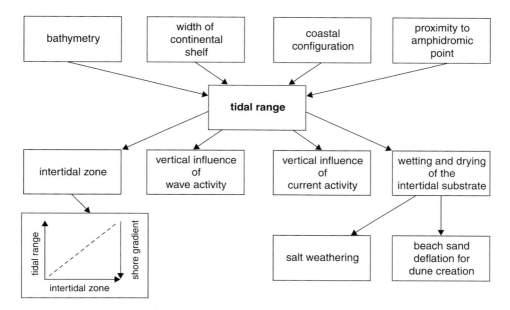

Figure 3.7 The geomorphological significance of tidal range.

may occur at and above the threshold velocity for a given particle size, but deposition occurs where velocity is less than the threshold. Figure 3.8b illustrates this with reference to mud- and sand-sized particles, and shows the relationship between current velocity and sediment size distribution within the intertidal zone, with mud characterising the low-energy low and high intertidal areas, whilst sand shoals occur in the high-energy mid-intertidal zone. Overall much sediment is transported in, out and within the intertidal zone during every ebb-flood tidal cycle (French, 2007).

Quite often a **sediment couplet** may be deposited (Plate 3.1), where sand deposited during tidal current deceleration is overlain by mud deposited at slackwater. If an additional couplet is deposited during each cycle then stacks of couplets may be preserved as **tidal**

(a)

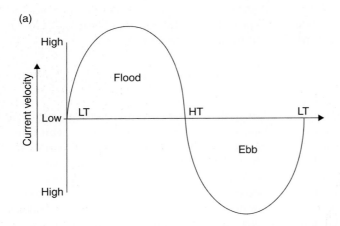

(b)

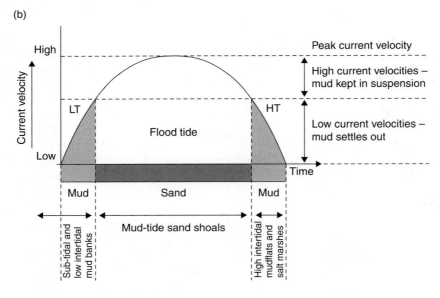

Figure 3.8 Tidal current activity and sediment deposition: (a) changes in current velocity and direction through a tidal cycle; and (b) the relationship between flood tide current velocity and the distribution of different sediment types in the intertidal zone (HT = high tide; LT = low tide).

Source: French (1997: 45, fig. 2.10).

Plate 3.1 A salt marsh cliff in the Baie de Mont St Michel (Normandy, France) showing distinct sediment couplets. Note coin for scale (*c.* 2 cm diameter) near the top of the cliff.

rhythmites. These have been used to reconstruct past tidal regimes, with the thickness of sand layers indicating spring to neap tide variations: the thicker sand layers represent the higher current velocity associated with spring tides. Indeed, if neap tide current velocities fail to exceed the sand transport threshold then the neap couplets will be dominated almost exclusively by mud (Reading and Collinson, 1996).

In many instances, due to factors similar to those discussed in section 3.2, the ebb-flood tidal cycle is asymmetrical, with either the flood or ebb tide taking longer than the other. For example, a 12-hour symmetrical tide consists of 5.5 hours per flood and ebb tide element, with 1 hour slackwater at high tide; however, an asymmetrical tide might comprise a flood tide lasting 3.5 hours and an ebb of 7.5 hours (French, 1997), with 1 hour at high water. This is of great significance for tidal geomorphology because a fixed volume of water or **tidal prism** must come in and go out each ebb-flood tidal cycle, which means that during the shortest tidal element, which in the above example is the flood tide, the tide must flow faster. The faster tidal velocity associated with the flood tide in the above example, means that more and coarser sediment can be transported during the flood, which becomes the **dominant current**. The ebb tide, however, becomes the **subordinate current**. A tidal coast may be referred to as flood or ebb dominated, depending upon which tidal element provides the dominant current. Extending this to sediment transport suggests that there is a net landward transport of sediment in flood-dominated systems (tidal inlets and estuaries may silt up as a consequence) and net seaward transport in ebb-dominated systems.

Scientific Box 3.2

Tidal sedimentary bedforms

The symmetry of the ebb-flood tidal cycle influences the types of sedimentary bedforms that occur within the intertidal zone. Sediment bedforms in tidal environments can be divided into two categories:

- **parallel orientated bedforms** are those with their long axis orientated in the same direction as the tidal current flow and include furrows, gravel waves and sand ribbons; and
- **perpendicular orientated bedforms** are those with their long axis at right angles to the tidal current flow direction and include dunes, sand waves and ripples; like aeolian sand dunes discussed in section 2.7.1 (Fig. 2.21), these bedforms may possess a cross-bedded structure that indicates the direction of the current that formed them.

The two categories tend to characterise higher and lower tidal current velocities respectively, but at both high and low velocity extremes a plane sediment surface may arise, unornamented by any bedforms. Under the influence of a symmetrical tidal regime, mobile sediment usually develops bedforms in accordance with the prevailing element of the ebb-flood tidal cycle, so that, for example, the lee side of a ripple will point shoreward on the flood and reorientate to seaward on the ebb. However, under an asymmetrical tidal regime the velocity of a subordinate current may not be sufficient to completely reverse a bedform created by the dominant current. For example, a sand wave or ripple created by a dominant current would be draped by mud at slackwater; some sediment remobilisation may occur during the subordinate current phase, particularly reworking ripple crests along what is termed a reactivation surface, although its original orientation would be preserved. Deposition of this reworked sand occurs on top of the reactivation surface and previous mud drape where it still persists. A second mud drape may then be deposited at slackwater on top of the thin reworked sand layer. Tidal couplets preserved in this situation would be cross-bedded alternating thick and thin sand layers or bundles separated by mud drapes (Reading and Collinson, 1996).

3.5 Estuaries

Estuaries are perhaps the best known of tide-dominated coastal systems. They are often semi-enclosed with a restricted opening to the sea, and commonly occur where the sea has invaded and drowned valleys and lowlands following the post-glacial rise in sea level. It is here that mixing between saline sea water and fresh river water occurs. Estuaries are host to a wealth of wildlife that exploit environments within the intertidal and subtidal zones, but estuaries are also important sites for human settlement and industry, and as such experience increasing environmental stress.

A large number of definitions for estuaries have been suggested over the last 50 years, but Dyer (1997: 6) has incorporated earlier definitions: 'An estuary is a semi-enclosed coastal body of water which has free connection to the open sea, extending into the river as far as the limit of tidal influence, and within which sea water is measurably diluted with fresh water derived from land drainage'. Although Dyer states that this is the most satisfactory overall definition, it is slightly ambiguous regarding river sections that are tidally influenced, but are beyond the upstream limit of sea water incursions. This definition acknowledges the importance of tides and salt-fresh water mixing in estuaries, but ignores sedimentological and biological aspects, which are included in definitions by Dalrymple *et al.* (1992) and Perillo (1995), respectively.

3.5.1 Estuary classification

A number of schemes have been proposed to classify estuaries, but most schemes are based on different aspects of the estuarine system, such as tidal range, topography, morphology, sedimentology, salt and fresh water mixing, and circulation (Dyer, 1997), and few attempt integrated classifications (e.g. Reinson, 1992). Some schemes employ complex quantitative techniques that are beyond the scope of an introductory text like this, and others consider a wider variety of aspects, such as biological and socio-economic factors (see Elliot and McLusky, 2002, for a discussion). The following sections concentrate on outlining the schemes that are useful in the field and in understanding sediment dynamics within the estuarine system.

3.5.2 Estuary morphology

The shape of estuaries is very varied, but all possess:

- an estuary head, where the river enters the estuary
- estuary margins that define the sides of the estuary
- an estuary mouth, where the estuary is open to the sea.

Following post-glacial sea-level rise many valleys have been inundated and drowned by the sea. An estuary occupying a former river valley is known as a **ria**, and will retain the meanders, tributaries and other features associated with the original watercourse. The same is true for a former glaciated valley, which when drowned becomes a **fjord**. In both rias and fjords the rate of relative sea-level rise outpaces the sediment infilling of the drowned valley, so that estuary morphology remains determined by the sides of the former valley and an **open-ended estuary** mouth persists. If sediment supply is high and the rate of deposition approaches being equal to the rate of sea-level rise then depositional features, such as salt marshes, occur around the estuary margins, narrowing the main tidal channel of the estuary. In many instances, where sediment is abundant within the estuarine system, a bar or spit may be built up by waves and longshore currents to partially obscure the estuary mouth to become a **bar-built** or **partially closed estuary** (see also section 2.6.3) (Fig. 3.9). Lagoons often occur just inside the mouth of a bar-built estuary, and occasionally the bar may extend to seal off the estuary mouth completely to form what is known as a **blind** or **lagoonal estuary**. Tidal current velocity is usually high in the constricted mouth of a bar-built estuary and coarse sediment brought into the estuary on these high-velocity currents during the flood tide will be deposited just inside the mouth as a **flood-tide delta** (see Plate 3.2). In temperate regions, extensive flood-tide deltas may develop during the summer months, often causing navigation problems within the estuary. However, these sediment build-ups are usually flushed out of the estuary during times of heavy river discharge associated with winter storms.

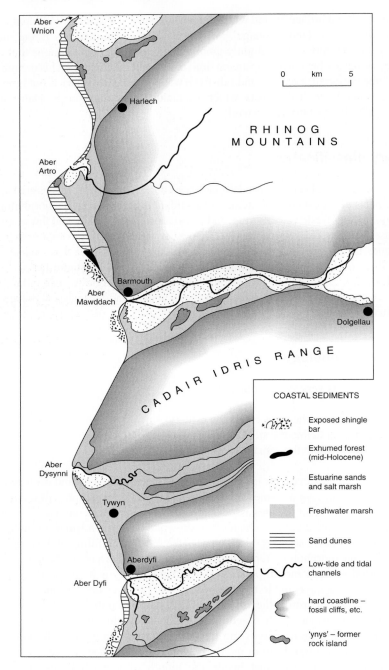

Figure 3.9 A series of bar-built estuaries on the coast of Cardigan Bay, Wales, UK.

Source: Briggs *et al.* (1997: 315, fig. 17.13).

Plate 3.2 A downstream view of a road embankment built across L'Aber Estuary (Brittany, France). Note the flood-tide delta on the upstream side of the embankment, and the extensive tidal flats accreting on the downstream side.

Management Box 3.3

Human interference in bar-built estuaries

Bar-built estuarine systems are particularly sensitive to human interference. An illustration of this is provided by ill-considered road building preferences in some regions. For example, in Brittany (northwest France) engineers have been inclined to construct embanked roads across the middle of bar-built estuaries, rather than bridges. Small fixed bridges are incorporated into embankments to allow river water to reach the sea, but these seldom constitute more than 3–4% of embankment length. Apart from restricting navigation, this has a number of environmental consequences:

1. The embankment acts like a dam, generally increasing upstream water levels. In some estuaries, such as L'Aber Estuary on the Crozon peninsula (Plate 3.2), where a barrage was built in 1958, this has caused water logging of adjacent lowland soils leading to the demise of the woodland it supported and general habitat change.
2. Flood tides flowing through the bridge deposit deltas on the upstream side, which can further reduce river discharge through the bridge.
3. Sediment introduced and deposited into the outer estuary, downstream of the embankment, is protected and no longer flushed out of the estuary during periods of heavy river discharge. In this way the estuary becomes a sediment

sink and continued sediment accretion leads to the progressive formation of tidal flats and salt marshes, and ultimately terrestrialisation.

4. The bars themselves are also affected, because these bars usually comprise dunes that rely on sediment flushed out of the estuary to supply and maintain the integrity of the dune system. With sediment trapped in the estuary, dune fronts become eroded and vulnerable. This situation applies to the bar dunes of L'Aber Estuary, and more seriously on the River Ster at Lesconil in southern Finistere where tidal flats and salt marshes behind the bar have been reclaimed and developed since 1850, and are now at risk due to the retreat of the bar dunes that protect the area. This is due to the construction of a barrage across the River Ster some time between 1952 and 1981 that has transformed the estuary into a sediment sink.

5. The residency time of pollutants within the estuary is increased, exacerbating the problem of eutrophication.

3.5.3 Salt and fresh water mixing within estuaries

The manner in which salt water of sea origin mixes with fresh water of river origin has implications for understanding estuarine sedimentology and provides a convenient means of classifying estuaries regardless of morphology. However, classifying estuaries according to mixing requires detailed measurements of the salinity structure of the estuarine water column, and is therefore not so readily utilised in the field without specialised equipment. Pritchard's (1955) seminal work on estuarine circulation has been the foundation for much subsequent research into estuarine hydrodynamics, and forms the basis of the classification described below.

Pritchard considered estuaries using a **salt-balance principle** that states mathematically that the rate of salinity change at a fixed point within an estuary is brought about by the operation of two processes:

1. **Diffusion** occurs when the difference in the ionic composition of salt and fresh water produces turbulence that mixes the water in the estuary to ultimately attain a uniform salinity.

2. **Advection** involves the physical mixing between salt and fresh water due to internal circulation.

Although these processes co-occur, one of them may dominate mixing in a given estuary. It is worth emphasising the difference, in that diffusion may be regarded as chemically driven mixing, whilst advection represents physical mixing. The manner and degree of mixing in an estuary determines its type according to Pritchard's classification (Fig. 3.10).

3.5.3.1 Stratified estuary

In estuaries where the mixing of salt and fresh water is minimal, the water column becomes stratified, with a lower high salinity layer and an upper fresh layer (Fig. 3.10a). The layer sequence is determined by density differences, with denser sea water occurring below the lighter and buoyant fresh water. Stratified estuaries are most common along protected microtidal coasts, and because of this the relative movement of the two flows, even during

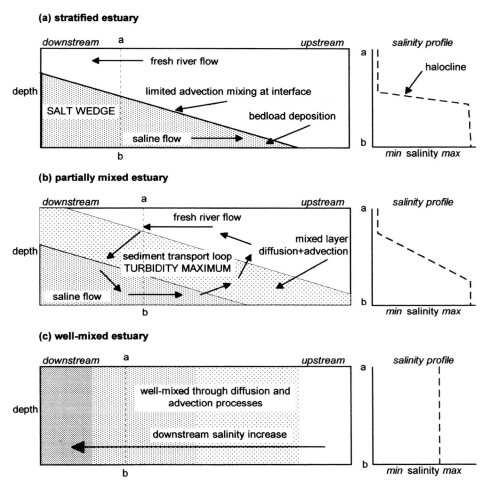

Figure 3.10 Estuary types according to Pritchard's (1955) salt-balance principle classification: (a) stratified estuary; (b) partially-mixed estuary; and (c) well-mixed estuary.

the ebb-tide phase, is almost always in opposite directions, so that the lower saline layer deforms to offer least resistance to the faster overflowing river water. In doing so, the saline layer thins or tapers upstream to form a **salt-wedge**. The interface between the two flows is known as the **halocline**, because it is here that salinity changes abruptly. As with all interacting fluids of differing densities (Fig. 2.7), waves form at the interface and these waves allow some limited physical mixing (i.e. advection) to occur along the halocline. Sediment behaviour within such an estuary is strongly influenced by the hydrodynamic conditions. Suspended sediment of river origin is often retained within the upper fresh layer until the open sea is reached, where a sediment plume extending out to sea may be observed. Sediment transported as bedload, however, whether of river or marine origin, is most often deposited at the tip of the salt-wedge, as it is here that flow velocity is either reduced or direction reversed depending on the state of the tide. The position of this bedload deposition shifts up and downstream in response to changes in tides and river discharge, the latter being the most important. During episodes of high river discharge, such as during floods, these bedload deposits may be flushed out of the estuary entirely. Fjords are also examples of stratified estuaries, but lack salt-wedges. Stratified estuaries have been variously referred to as type A, river-dominated, salt-wedge (or fjord) estuaries. A well-known example of a stratified estuary is that of the Mississippi River in the USA.

3.5.3.2 Partially-mixed estuary

These are more influenced by tides than stratified estuaries and are typical of mesotidal to macrotidal settings. Tidal turbulence, caused by the ebb and flood entering and exiting the estuary, destroys the interface between the salt-wedge and overlying fresh water to produce a more gradual salinity gradient through the water column (Fig. 3.10b). Both advection and diffusion processes operate, and large density-driven eddies exist which help to exchange salt water upwards and fresh water downwards. A broad mixed salinity zone is created, corresponding to the steepest part of the vertical salinity gradient (i.e. halocline), which shallows downstream, so that it is at the bed near the estuary head and at the water surface near the estuary mouth. This inclined mixed zone, therefore, is slightly reminiscent of the salt-wedge morphology, occurring between the opposing river and tidal flows. Thus, a two-layer flow exists, separated by a **level of no motion** where the average flow velocity is zero, and which approximately corresponds to both the position of the mixed zone and the halocline. Also, the location on the estuary bed where the tidal and river flows meet and converge is known as the **null point** (Dyer, 1997).

The sediment dynamics of a partially-mixed estuary are substantially different from those in a stratified type. Suspended sediment transported downstream in the river flow will at some point encounter the mixed zone and its associated lower flow velocity. Particles will begin to settle out, but once below the mixed zone they become entrained and transported back upstream by the tidal flow. The same particles may then re-encounter the mixed zone and be mixed upwards by eddies into the river flow. Particles may circulate in a closed loop like this for some time, and the high concentration of suspended sediment trapped in this way around the mixed zone is known as a **turbidity maximum**. The position of the turbidity maximum moves up and down the estuary with the ebb and flood of the tide. Partially-mixed estuaries have also been referred to as type B estuaries, and a good example is the Mersey in northwest England.

3.5.3.3 Well-mixed estuary

This estuary type is dominated by tidal activity and requires severely macrotidal and hypertidal conditions to effectively mix the waters, through both advection and diffusion processes. A well-mixed estuary, or vertically homogeneous estuary, characteristically lacks a vertical salinity gradient, so that salinity is uniform from surface to bed at any given point within the estuary (Fig. 3.10c). However, there are three subdivisions within this category.

1. Where the estuary is particularly wide the Coriolis effect may separate the flows so that the seaward river flow is restricted to the right side of the estuary (in the northern hemisphere) and the landward tidal flow to the left, but within each flow salinity remains uniform throughout the water column. This type has been referred to as a **laterally inhomogeneous estuary**.

2. Some estuaries have no separation of river and tidal flows, and in such cases sufficient mixing may give rise to uniform salinity at all points across an estuary, so producing a **laterally homogeneous estuary** (also known as a sectionally homogeneous or type C estuary by some authors). However, salinity does change with distance down the estuary, so that minimum salinity occurs at the head and maximum salinity towards the mouth of the estuary.

3. Where salinity is uniform both laterally across an estuary and longitudinally along the length of an estuary, from head to mouth, then a truly **homogeneous estuary** (or type D estuary) is defined. This type constitutes the theoretical end member of the spectrum of estuary types, and opposes the highly stratified estuary described above.

3.5.4 Estuarine sedimentation

The sediment within an estuary comes from one of three main sources:

1. *Fluvial/glacial sources.* This is sediment supplied by the river(s) flowing into an estuary, and/or from glaciers along cold coasts. It is considered to be of terrestrial origin and has generally increased through historic times with the advent of widespread farming, which has stimulated soil erosion in many river catchments. The supply of this sediment may be seasonal, often being associated with high precipitation events during winter storms.
2. *Estuary margin sources.* This describes sediment eroded from the margins of the estuary itself. The material being eroded may either be soft sediment previously deposited by the estuary, or a geologically older material, such as the local bedrock. Tidal currents working within the estuary may be sufficient to rework soft sediment, but erosion increases with occasional high wave-energy activity within the estuary. Indeed, significant erosion of hard bedrock requires these increased energy levels.
3. *Extra-estuary sources.* This encompasses sediment supplied from outside the estuary mouth, and includes sediment eroded from cliffs along the coast downdrift of the estuary mouth, and continental shelf sediment thrown into suspension by passing waves. Near the estuary mouth this material is entrained by flood tidal currents and transported into the estuary.

Once the sediment is supplied to the estuary from one of the various sources, its fate is determined by the combined effects of sediment particle size, wave-tide-river energy, and the salt-fresh water mixing relationship within the estuary (Fig. 3.11a). Sediment composition in an estuary that is dominated by fairly coarse sand-grade material is known as a **non-turbid estuary**, so-named because there is a low concentration of **suspended particulate matter** (SPM) within the water column. The sand may be transported in suspension during periods of maximum current velocity, but quickly settles out following the peak tidal flow, leaving the water relatively clear (i.e. non-turbid). A **turbid estuary**, however, is characterised by fine sediment composition of mud- to silt-grade. These particles are easily held in suspension and so the concentration of suspended particulate matter is usually high at all states of the tide. The deposition of these fine particles is aided by the process of **flocculation**, which occurs when clay particles in the river water come into contact with salt nuclei upon mixing in the estuary. This process encourages clay particles to join together or coagulate to form larger particles called **flocs**, which are heavier and so more readily deposited. Fine sediment may also be ingested by invertebrates and exported to the estuary bed as faecal pellets. Carter (1988) describes the behaviour of suspended sediment in a turbid estuary through a tidal cycle. He suggests that at peak tidal flow sediment throughout the water column is well mixed, producing a uniform SPM concentration at all depths. As the tidal flow decelerates SPM starts to become more concentrated at depth, producing a sediment concentration gradient with depth, known as the **lutocline**. With progressive deceleration of the tidal flow the lutocline may become stepped or stratified, until finally these layers merge at or above the bed at the time of slackwater. The sequence is then reversed as the tidal flow accelerates following slackwater.

(a)

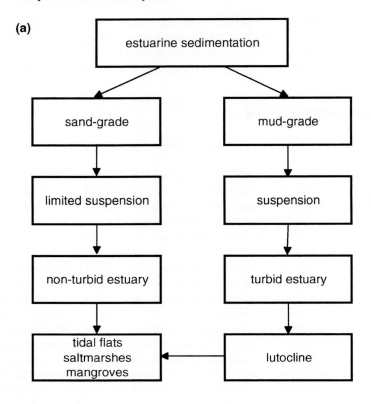

(b)

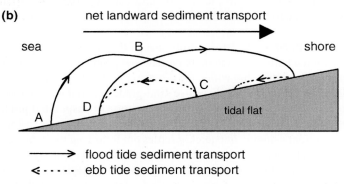

Figure 3.11 (a) Sediment pathways in an estuarine system. (b) The mechanics of settling lag (see text for explanation) and its role in inducing net landward sediment transport.

Sedimentary environments and particle-size gradients within an estuary have been described by Dalrymple *et al.* (1992). They divide estuaries longitudinally, according to energy sources and their relative amounts, into a marine-dominated outer section near the mouth, a mixed-energy central section around the estuary mid-point, and a river-dominated inner section near the head. The outer and inner sections have high energy levels related to marine and riverine input, respectively, whilst the central mixed-energy section generally possesses the lowest energy. This is because marine and riverine energies decrease up- and down-estuary, respectively, both being relatively low in the central section. The sedimentological consequence of this energy partitioning is a saddle-shaped particle-size gradient along the estuary, with relatively coarse sediment occurring at both the high-energy head and mouth regions, and finer sediment in the lower-energy central section. Dalrymple *et al.* also suggest an estuary classification based on the dominant marine process, so that they recognise distinct **wave-dominated** and **tide-dominated estuaries**. Reinson (1992) states that wave-dominated estuaries can occur in microtidal to low macrotidal settings of stratified to partially-mixed type, and morphologically include blind and bar-built to open-ended estuaries. Tide-dominated estuaries, however, occur only under high macrotidal to hypertidal conditions, associated with well-mixed and open-ended estuary types.

In addition to particle size variation along an estuary, size variation also occurs laterally across an estuary. Energy tends to decrease from the central estuary axis to the intertidal margins, and this corresponds to a similar decrease in particle size. Fine sediment tends to be transported landward and deposited close to the margins of the estuary due to the phenomenon known as **settling lag** (Postma, 1961), which contributes to the formation of depositional features such as mudflats, salt marshes and mangroves. Settling lag works when a flood tidal current attains a threshold velocity sufficient to entrain a given sediment particle at location A, which is subsequently transported landward in the flow. The velocity of a flood-tide current decreases landward, in addition to the approach of slackwater, so that the velocity eventually falls to the entrained particle's **settling velocity** threshold at B. The particle does not fall vertically to the sediment surface at B, but is carried by inertia further landward to C. The particle is re-entrained by the ebb at C, but at a much later stage in the ebb cycle. Therefore, the particle is transported seaward for a shorter time period and is deposited at D, which is landward of its original position A (Fig. 3.11b).

Tidal sand and mudflats may develop then as a direct consequence of settling lag, eventually building up to allow vegetation to take hold to become salt marsh or mangrove (see section 3.5). However, in arid environments a particular type of tidal flat environment may develop, called a **sabka**. This is a distinct coastal environment that characterises arid regions where it replaces the more familiar mudflats, salt marsh and mangroves (Box 3.4).

Scientific Box 3.4

Sabkas – tidal flats of arid environments

These are low gradient salt flats found in arid environments (Mather, 2007), with extensive examples in the Arabian Gulf, Egypt and Mexico, and they occur both in the intertidal zone and supratidally. Mangroves are sometimes associated with sabkas, but most frequently they support algal mat communities that are tolerant of the arid and saline conditions. Salts often precipitate to form hard salt crusts on the surface, and other evaporite minerals are also common, such as gypsum (Briere, 2000). Both clastic and carbonate sabkas exist, the former being associated with deltaic sedimentation, such as at Bahrah in northern Kuwait, part of the Tigris–Euphrates Delta (Saleh *et al.*, 1999). The Bahrah sabka is approximately 5.5 km wide and is apparently unique in that is comprises a landward sediment zone characterised by terrigenous quartz silt and sand deposited through aeolian processes, and the more usual seaward carbonate and evaporite deposits. The development of the extensive Bahrah sabka has taken place since 3040 years ago, when the sea levels of the region stabilised, growing seaward at a rate of 1.5 to 2 m per year.

3.5.5   Estuarine ecology

With the exception of high intertidal salt marshes and mangroves, which are discussed in section 3.6, estuaries typically lack abundant macroflora. This is because the subtidal and low intertidal sediment substrate is highly mobile, and only occasional stones or gravel patches will support algal colonies. Higher up in the intertidal zone the environment is less energetic and as a consequence the sediment is less mobile. Under such conditions, oxygen is depleted rapidly within the sediment profile leading to anoxic conditions that support hydrogen-sulphide-producing bacteria. This results in pungent, thick black mud being created just under the active sediment surface. A rich community of microflora and microfauna, such as diatoms and foraminifera, exists in this environment, as do a number of burrowing higher-order organisms, such as ragworms and various molluscs. This infauna in turn supports diverse and abundant populations of fish and wading birds.

A significant environmental parameter to affect estuaries, more so than most other coastal environments, is salinity. Salinity can range from fresh water to normal-salinity sea water of approximately 33‰ salt content. Although any given point in an estuary experiences a range of salinities throughout the tidal cycles, the longitudinal salinity gradient down an estuary produces a marked ecological zonation, particularly apparent in invertebrates (Cremona, 1988).

3.6 Salt marshes and mangroves

For most of the tidal cycle, the upper part of the intertidal zone is exposed to subaerial conditions, and it is here in mainly tide-dominated situations that salt-tolerant plants may grow to create widespread vegetated intertidal surfaces, known as either salt marshes or mangroves. They also develop on deltas, but these are mainly river-dominated and will be discussed in the next chapter. Salt marshes are characterised by short plants, such as grasses, and are mostly restricted to temperate coastlines (Allen, 2000a), whilst mangroves are their tropical and subtropical counterparts, and comprise trees of various heights that can develop into extensive mangrove forests or mangals (Perry, 2007). The vegetation is often zoned, from a zone of pioneer species seaward, grading landward to a more mature community. In salt marshes, these vegetation changes are used to subdivide the environment into low and high marsh, respectively, although these again may be separated by an additional intermediate or middle marsh zone. Mangroves in tide-dominated settings may also display a vegetation zonation. Both have common aspects to their geomorphology and sedimentology, but distinct differences make separate treatment here logical.

3.6.1   Salt marsh geomorphology

The large-scale geomorphological impression of these coastal systems is that of a near- or quasi-horizontal platform, that slopes gently seaward. However, on smaller scales a wide variety of features may be seen, such as cliffs, salt pans, tidal creeks and their levées. The platforms are built up by the deposition of sediment being brought onto the marsh surface by flood-tide currents and trapped by vegetation. In this way, sediment **accretion** on the surface of a marsh leads to its elevation within the tidal frame; it is worth noting that the rate of surface elevation is usually less than accretion, because accreted sediment often compacts under the weight of successive episodes of deposition. As we will see in the next section, the rate of accretion varies across a marsh transect, so that relatively more sediment is accreted lower in the tidal frame than near the high-tide limit. This process ensures that lower surfaces elevate at a faster rate, so as to catch up the initially higher surfaces, resulting in a quasi-horizontal platform. Salt marshes may exist on a wide variety of tidally influenced coastlines. French (1997) illustrates seven topographically distinct salt marsh settings, including those found along open coasts and embayments, those protected behind barriers, and those that occur along the fringes of estuaries (Fig. 3.12).

Features marking the boundary between marsh platforms and seaward mudflats are variable. Allen (1993, 2000a) documents three major categories of marsh-mudflat boundaries that he recognises along turbid British tidal coasts, although the categories can be widely applied (Fig. 3.13):

1. *Ramped marsh shore.* This is a gently sloping, smooth transition from mudflat to marsh, and may exist where sediment supply is abundant. In these conditions, the marsh is actively accreting seaward and indicates a positive shore regime.
2. *Cliffed marsh shore.* Under normal circumstances very little vertical erosion of sediment occurs from the marsh/mangrove surface, but this does not include sediment resuspended from the layer at the base of the lutocline at slackwater, which fails to become deposited. Quite often, however, cliffs up to several metres high occur at the edge of salt marshes, commonly separating the marshes from seaward mudflats. These cliffs are evidence that erosion is taking place in the horizontal plane and that the marsh is retreating landward, indicating a negative shore regime. Allen (1989) documents the principal failure mechanisms for marsh cliffs, which include toppling failure and rotational slumping (see Fig. 2.10). Erosion and cliff development are associated with the meandering of tidal channels, increases in wave activity, and decreases in sediment supply, although once initiated, wetting-drying cycles and seasonal changes in climate help to perpetuate cliff retreat (Plate 3.3).
3. *Spur and furrow marsh shore.* This is the least common of the shore types, although it may be locally widespread. It is characterised by a generally ramped shoreline that is dissected by numerous furrows, which are separated by ridges or spurs. The furrows are products of very localised erosion, which upon initiation trap pebbles, gravel and sand which get washed up and down the furrow by the tide, thus aiding further erosion and furrow deepening. Deposition occurs on top of the spurs, and in this way furrow erosion and spur-top deposition are balanced, with neither retreat nor advance dominating, indicating a neutral shore regime.

Tidal creeks dissect salt marshes and are the principal conduits through which tidal water and sediment floods onto and ebbs off the salt marsh surface. Morphologically, tidal creeks are similar to river channels in that they often develop dendritic networks and meandering channels complete with levées. However, the hydraulic and flow conditions differ from rivers, because whereas rivers are at their most energetic when the channel is full (i.e. under storm conditions), tidal creeks are at their most sluggish at bank full, because this usually occurs at slackwater associated with high tides, when flood flows reverse to become ebb flows. Under fairweather conditions, the ebb tide appears to be dominant in creek channel and headwater erosion. Commonly a cascade of smaller channels may be seen embedded in a creek cross-section, each representing erosion during later stages in the ebb cycle under decreasing discharge (Plate 3.4). However, during storm conditions, the flood tide may become dominant, with high flow velocity and discharge directed landwards within the creeks (Bayliss-Smith *et al.*, 1979; Knighton *et al.*, 1992).

Throughout the development of a salt marsh, the characteristics of tidal creeks change in response to their changing **hydraulic duty**. Under periods of rapid sea-level rise, salt marshes are expected to be positioned low within the tidal frame so large creeks would be needed to convey a greater tidal prism on and off the marsh, whereas creek size would

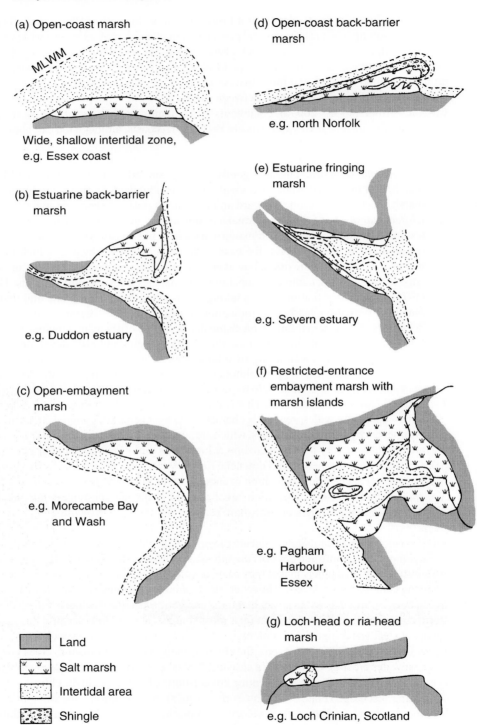

Figure 3.12 Various salt marsh settings in tidal environments.

Source: French (1997: 46, fig. 2.11).

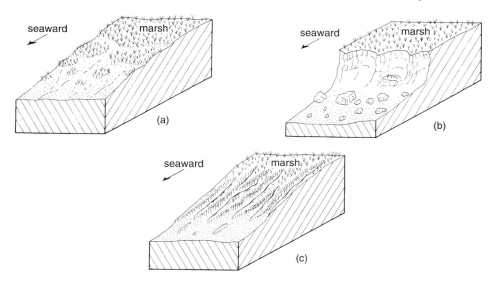

Figure 3.13 Three categories of salt marsh shorelines: (a) ramped; (b) cliffed; and (c) spur and furrow.

Source: Allen (1993), published with the permission of the Geologists' Association.

Plate 3.3 Extensive cliffing of a salt marsh at Northwick Oaze in the Severn Estuary (UK). The sediment corer is *c*. 1.25 m long.

Plate 3.4 A nest of channels in a salt marsh tidal creek at Barneville (Normandy, France). The salt marsh vegetation is mainly sea purselane (*Halimione portulacoides*).

decrease under lower rates of sea-level rise. In areas where preservation is good, tidal creek palaeochannels of various sizes may occur nested within the stratigraphic sequence, as seen in the Severn Estuary in the UK (Allen *et al.*, 2006), where they reflect changing environmental conditions through the Holocene.

Salt pans are unvegetated pools of standing water that occur on the salt marsh surface. There are two main morphological types: roughly circular 'primary' pans, and elongate channel pans. Some controversy surrounds the origin of 'primary' pans, as some authors consider them to develop from the outset of salt marsh formation (hence 'primary' pans) as an expression of uneven accretion on the surface, so that areas of low accretion develop into pans. Others have suggested that they are formed by tidal litter scouring hollows that develop into pans. The origin of channel pans, however, appears more secure in that they represent tidal creeks that have been severed by fallen sediment, creating a dam which ponds up ebb tidal water. Whatever their origin, salt pans are conspicuous features of most salt marsh systems, and play an important ecological role.

3.6.2   Salt marsh sedimentology

A number of models have been proposed to describe the sedimentation on salt marshes. All are relatively similar, often employing daunting expressions. Allen (1994) proposed such a model, but thankfully later summarised it in a more elementary form (Allen, 1996). Although the details of the model are not appropriate here, the results of the model are useful as aids to understanding general principles of salt marsh sedimentation and subsequent development (Fig. 3.14):

● The model simplifies the real world by assuming a constant morphology for the salt marsh shore, in that a horizontal salt marsh platform is bounded to the landward side by a barrier, such as an artificial sea wall or natural cliff, and to the seaward side by a tidal channel, which could be a tidal river in an estuarine setting. The tidal channel is the source of both water and sediment delivered onto the marsh platform.

● The model predicts that at any time when the salt marsh surface is submerged, the velocity of the tidal flow decreases linearly landward, whether associated with the flood or ebb tides. The model takes no account of friction or turbulence, so that the contrast between the high velocity channel and low velocity platform is solely due to differences in the momentum, competence and capacity of the two environments.

● The landward decrease in flow velocity has three major sedimentological conse-quences: (1) reduced flow velocity allows relatively coarse sediment to be deposited shortly after entering onto the marsh platform, so that the model predicts a landward decrease in sediment grain size across the marsh; (2) this rapid deposition of coarse sediment near the marsh edge contributes to a predicted overall landward decrease in deposition rate; and (3) the sediment concentration of the tidal water across the marsh is also predicted to decrease landward, a consequence which is closely related to the first two points.

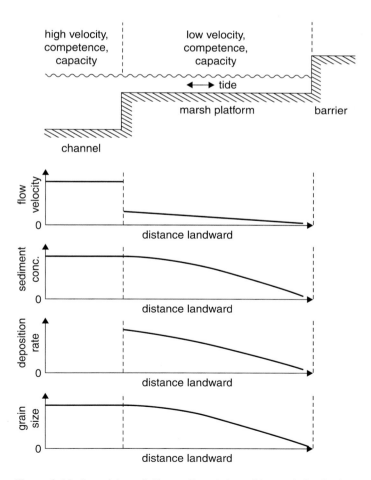

Figure 3.14 A model predicting sediment deposition and distribution upon quasi-horizontal salt marsh surfaces.

Source: Allen (1996), published with the permission of the Geologists' Association.

3.6.3   Salt marsh morphodynamics

Like most coastal sedimentary systems, salt marshes are extremely sensitive to changing environmental conditions. Changes in sediment accretion and erosive processes are amongst the principal controls on salt marsh morphodynamics, and in particular in determining the position of salt marsh shorelines. Figure 3.15 shows the relationship between these variables and indicates that during periods when the influence of erosion is greater/less than the horizontal accretion of sediment, the marsh shoreline will retreat/advance (negative/positive shore regimes). Where the regime alternates between negative and positive, then successive phases of erosion and accretion may occur. These phases manifest themselves as series of seaward-descending terraces, each separated by small clifflets, which are the protruding remnants of cliffs developed in the retreating phases, and then partially buried by subsequent accretion. Allen (1993) describes these as offlapping morphostratigraphic units, and notes their widespread occurrence in estuarine fringing marshes (Fig. 3.16). In the hypertidal Severn Estuary in southwest Britain, at

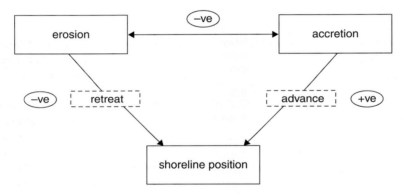

Figure 3.15 The relationship between horizontal erosion and accretion in determining salt marsh shore regime. Where accretion > erosion then the shore will advance (positive shore regime); where accretion < erosion the shore will retreat (negative shore regime).

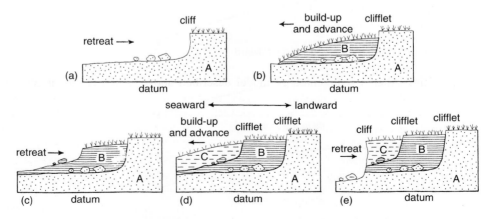

Figure 3.16 Alternating negative and positive salt marsh shore regimes resulting in the formation of seaward-descending offlapping morphostratigraphic units.

Source: Allen (1993), published with the permission of the Geologists' Association.

least three major morphostratigraphic salt marsh units have been recognised and given full geological formation status. The oldest is the Rumney Formation which began accreting in the seventeenth century, then the eighteenth century Awre Formation, and finally the Northwick Formation which started forming in the twentieth century. A period of horizontal erosion and shoreline retreat stratigraphically separates each formation, but each formation continues to vertically accrete sediment from the outset.

Surveys of older salt marsh sediments throughout the Holocene and historic period show variation in morphostratigraphic sequences (e.g. Allen and Haslett, 2002, 2007). Allen (2003) models these variations in terms of the response of salt marsh sedimentation to changes in sea level. He recognises that sequences may be deposited: (1) continuously, as sea level fluctuates around a general rising trend resulting in alternating silt and peat layers, dependent upon whether the rate of sea-level rise was higher or lower, respectively; or (2) where erosion of a salt marsh produces an erosion surface that is subsequently buried by rapid salt marsh sedimentation, often being characterised by sand laminations reflecting seasonal changes in the depositional environment (Allen and Haslett, 2006).

3.6.4   Salt marsh ecology

Salt marsh geomorphology is closely linked to ecology, because it is widely accepted that salt marsh plants make a significant contribution to the trapping and accretion of sediment. They also produce large amounts of particulate and dissolved organic matter which play a part in coastal food web dynamics. Salt marshes first develop when unvegetated mud or sand flats become colonised by pioneering **halophytes**, which include plants such as *Spartina* (cord grass) and *Salicornia* (glasswort) species. This low salt marsh community has to withstand extreme environmental pressures, including the instability of the sediment and its high salt content, regular tidal submergence with its associated waterlogging effects, and minimal oxygen availability, as only the sediment surface is aerated. The upper salt marsh sediment is more stable, less frequently inundated by the tide, and so is able to support a more diverse plant community. In Europe, typical upper salt marsh plants include *Aster tripolium* (sea-aster), *Plantago maritima* (sea-plantain), *Limonium vulgare*

Technical Box 3.5

Reconstructing salt marsh shoreline position

Using Allen's sedimentology model described above, Allen (1996) is able to reconstruct historical variations in the shoreline position of salt marshes in the Severn Estuary. This is possible because at any fixed location on a salt marsh, particle size will increase through time as the shoreline retreats landward, and decrease as the shoreline advances seaward (Fig. 3.17). Therefore, according to the model, particle size variations in a salt marsh sediment sequence reflect changes in the position of the shoreline. This model is therefore, potentially very useful in investigating the extent and rates of salt marsh morphodynamic changes (Allen, 2000a). Indeed, Haslett *et al.* (2003) employed the model predictions to aid analysis of historic salt marsh sedimentation in Normandy (France), where they investigated its relationship to sea-level rise and sediment supply.

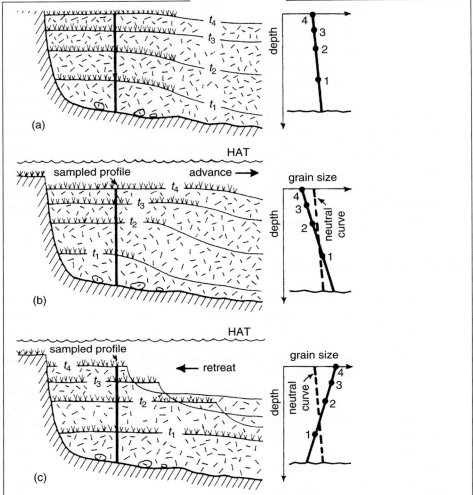

Figure 3.17 The influence of salt marsh shoreline position on temporal grain size distribution (HAT = highest astronomical tide).

Source: Allen (1996), published with the permission of the Geologists' Association.

(sea-lavender), *Cochlearia officinalis* (scurvey grass), *Halimione portulacoides* (sea-purselane) and *Puccinellia maritima* (salt marsh grass). Also, close to the landward limit of tidal inundation, the strandline is characterised by plants less tolerant of salt, such as *Juncus maritimus* (sea-rush) and *Phragmites communis* (common reed). Salt marshes, then, possess a clear plant zonation that reflects vegetation colonisation and succession, and is strongly influenced by factors associated with the frequency and duration of tidal inundation. Similar zonations are also exhibited by other organisms, such as diatoms and foraminifera. Animals that are found on salt marshes are typically terrestrial species at the limit of their range. However, the more marine lower salt marsh may be home to some marine animals, such as estuarine crabs (e.g. *Carcinus maenas*) and snails (e.g. *Hydrobia* species). Furthermore, because salt marshes are productive throughout the year, they are often important wintering sites for wildfowl (Cremona, 1988).

3.6.5 Mangroves

The principal difference between salt marshes and mangroves is the greater above-ground biomass in mangroves. There are many different species of mangrove trees, but amongst the commonest genera seen at a mangrove shore are *Rhizophora*, which possess dense networks of prop roots that extend into the sediment surface from the above ground trunk, and *Avicennia* that have a horizontal below-ground root system that sends up shoots called pneumatophores (Plate 3.5). Both strategies assist in anchoring the trees to the substrate, and also the partial above-ground roots help to facilitate oxygen intake, as the substrate is invariably anaerobic. Such root networks are also considered to promote the trapping and deposition of sediment that is introduced into the mangal system, up to approximately 2 mm/year.

Salt marshes and mangroves are geographically distinct, with salt marshes dominating in the temperate zones, whilst mangroves are restricted to lower latitudes between 32°N (Bermuda) and 38°S (Australia). Zonation of mangrove species does occur, with *Avicennia* and *Rhizophora* often characterising the seaward zones, whilst further inland species of other genera such as *Laguncularia* appear. Three main physical mangrove settings occur (Woodroffe, 1993):

1. *River-dominated setting* – refers to substantial mangroves that exist on many low-latitude deltas, where the sediment is principally supplied by river water; for example, the Fly River Delta in Papua New Guinea.
2. *Tide-dominated setting* – mainly associated with estuarine situations in which sediment is supplied by tides and tidal currents; for example, the South Alligator River in the Northern Territory of Australia.
3. *Carbonate setting* – restricted to coral reef settings where mangroves may grow around cays and in protected areas behind storm shingle ramparts and coral islands. The sediment here is of local derivation being supplied by wave erosion of coral. However, if this supply is limited the mangroves may develop organic-rich peat deposits. An example is Grand Cayman.

Mangroves are ecologically rich and are refuges for many species of birds and other animals. Also, the sheltered mangrove environment, particularly creek networks, are known to act as nurseries for a number of fish and sea-food (shrimp and prawn) species, some of which are important food and profitable species for local communities. Furthermore, dense mangals offer significant protection to coastal communities from tropical storm waves and associated erosion. However, this fact is often neglected and many mangrove sites in developing countries are being converted to agriculture, such as rice cultivation, which is ironically less profitable than harvesting shrimp in undamaged mangroves. For example, in some west African states original mangroves have declined in area by 46%, and in the Philippines by as much as 75% (French, 1997).

Summary points

- Tides are generated by the combined gravitational attraction of both the moon and the sun upon the earth, with different orbital configurations giving rise to spring and neap tides. Predicted tidal levels can be significantly altered by meteorological effects.
- Tidal range strongly influences the type of landforms present in a tidal coastal system, as it determines the spatial extent of marine activity at the coast, and influences the type and operation of weathering, erosional and depositional processes.

Plate 3.5 Mangroves of the eastern Australian coastline: (a) *Avicennia* mangroves with abundant pneumatophores (Minnamurra Estuary, New South Wales); (b) *Rhizophora* with prop root networks (Thomatis Creek, Queensland).

● Tidal environments, such as estuaries, salt marshes, mangroves and sabkas, possess distinctive ecosystems that are often under pressure from human society. For example, many estuaries are centres for urban settlement and industry.

Discussion questions

1. Evaluate the geomorphological significance of tidal range.
2. Assess the suitability of salt-fresh water mixing for the classification of estuarine systems.
3. Salt marshes are dynamic tidal coastal systems. Examine the dynamic relationship between the geomorphology and sedimentology of a temperate salt marsh.

Further reading

See also

Tide-dominated deltas, section 4.2.2
Coastal responses to sea-level change, section 5.3
Coastal management issues, section 6.2

Introductory reading

Tides, Surges and Mean Sea-Level. D. T. Pugh. 1987. Wiley, Chichester, 472pp.
Although over 20 years old, this book still remains a definitive and detailed account of tides, tidal predictions, levels and meteorological effects.

Estuaries: A Physical Introduction (2nd edn). K. R. Dyer. 1997. Wiley, Chichester, 210pp.
An introductory text to estuaries, including their classification, tidal effects and mixing processes, but quite mathematical in places.

The Estuarine Ecosystem (2nd edn). D. S. McLusky. 1989. Blackie, Glasgow, 215pp.
A now standard and well-accepted introduction to estuarine ecosystems.

Coastal and Estuarine Management. P. W. French. 1997. Routledge, London, 251pp.
Although aimed at being a general coastal management text, because of the author's interests it leans toward the management of tidal environments, especially estuaries.

Intertidal Ecology. D. Raffaelli and S. Hawkins. 1996. Chapman and Hall, London, 356pp.
A timely general introduction to the ecology of intertidal environments along all types of coasts.

Advanced reading

Estuarine physical processes research: some recent studies and progress. R. J. Uncles. 2002. *Estuarine, Coastal and Shelf Science*, **55**, 829–856.
An essay from a respected author in the field that examines physical estuarine processes.

High Resolution Morphodynamics and Sedimentary Evolution of Estuaries. D. M. Fitzgerald and J. Knight (eds). 2005. Springer, Berlin, 364pp.
A collection of papers that examine the relationships between sediments and estuarine landforms.

Estuarine Shores: Evolution, Environments and Human Alterations. K. F. Nordstrom and C. T. Roman (eds). 1996. Wiley, New York, 510pp.
An advanced compilation of work on the very varied nature of estuarine environments.

Estuaries: Monitoring and Modelling the Physical System. J. Hardisty. 2007. Blackwell, Oxford, 157pp.
A useful instruction manual to creating an estuarine model, with the Humber Estuary, UK, used as an example.

Saltmarshes: Morphodynamics, Conservation and Engineering Significance. J. R. L. Allen and K. Pye (eds). 1992. Cambridge University Press, Cambridge.
A useful and wide-ranging introduction to salt marsh systems, their ecology and applied issues.

Coastal and Estuarine Environments: Sedimentology, Geomorphology and Geoarchaeology. K. Pye and J. R. L. Allen (eds). 2000. Geological Society, London, Special Publication No. 175, 470pp.
A collection of advanced research papers on sedimentary, and especially estuarine, coasts.

Academic journals, such as *Estuaries, Estuaries and Coasts, and Estuarine, Coastal and Shelf Science*, are worth consulting regularly for current science in this area.

4 River-dominated coastal systems

River-dominated coastal systems are known as deltas, which are coastal accumulations of river-supplied sediment, although they also have complex interactions with waves and tides. They are often very large and act as centres for human settlement, agriculture and industry, but are also very sensitive to human interference. This chapter covers:

- the classification of deltas and the role of rivers, waves and tides
- the hydrodynamic relationship between rivers and the sea within the delta system
- the sedimentological and geomorphological characteristics of deltas
- the human habitatation of delta systems and their sensitivity to interference

4.1 Introduction

Deltas comprise many of the different coastal environment types that are associated with wave-dominated and tide-dominated systems, but their distinctiveness comes from the input of water and sediment from riverine sources. They are major physical constructions that occur at the coast, and are extremely important as locations for human settlement, agriculture and industry. Furthermore, they are dynamic systems, not only responding to changes in the river systems that supply water and sediment, but also to coastal processes: wave, tide and aeolian processes. They are fascinating coastal environments, often of regional if not global significance, and worthy of inclusion here.

The physical nature of deltas has been mainly established, not by coastal geomorphologists but by sedimentologists working in the petroleum exploration industry. This is because deltaic environments, mostly ancient, contain many of the sedimentological criteria necessary for hydrocarbon (gas and oil) development. This includes organic deposits, such as peat and coal that may give rise to hydrocarbons (source rocks), and also vast sand bodies with abundant and sizeable pore spaces between the sand grains, playing host to hydrocarbons (reservoir rocks).

The ecology of deltaic systems is not covered in this chapter. This is because deltaic ecosystems are extremely varied, from riverine habitats (which aren't coastal) to estuaries, mangroves, salt marshes, sabkas, barrier islands, coastal sand dunes and beaches, all of which are covered elsewhere in this book. Therefore, this chapter concentrates upon the hydrodynamics, geomorphology, sedimentology and human influence in the delta system.

4.2 Delta classification

Traditionally, deltas have been categorised according to their overall plan morphology (Table 4.1). Since Galloway (1975), however, deltas have been classified according to the

Table 4.1 Traditional categorisation of deltas according to their overall plan morphology

Delta plan	Description	Example
arcuate (fan-shaped) delta	triangular-shaped delta, with downstream branching distributary channels to the sea	Nile Delta (Egypt)
bird's foot delta	sediment builds up as levées either side of a channel, but deep water exists between channels	Mississippi Delta (USA)
compound delta	deltas merging from more than one river outflow	Ganges–Brahmaputra Delta (Bay of Bengal)
estuarine delta	drowned river valley infilled by the deposition of river sediment	Mobile River Delta (USA)
cuspate delta	a delta whose shoreline is extensively modified by wave action	Rhône Delta (France)

dominant process operating in an individual delta system (Fig. 4.1), so that a delta may be described as river-dominated, tide-dominated or wave-dominated. These process relationships are illustrated by a ternary model (Fig. 4.2) – a triangular diagram with the three processes each occupying a corner of the triangle. Individual deltas are then plotted within the diagram according to the relative importance of the different processes.

4.2.1 River-dominated deltas

These extend outwards from the coast due to sediment-laden river water jetting out into the sea. This tends to form levées that define the river channels, and hence the name **bird's-foot delta** has been applied to this rather spindly morphology. The Mississippi Delta in the southeast USA is a good example.

The significance of riverine activity in delta dynamics has been highlighted by Hensel *et al.* (1999) in a study of Rhône Delta along the Mediterranean coast of France. Sediment accretion and the resulting elevation of the delta surface has been measured between 1992

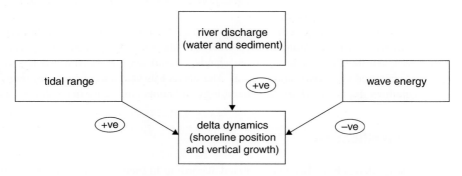

Figure 4.1 The relationship between delta dynamics and river, wave and tidal influences.

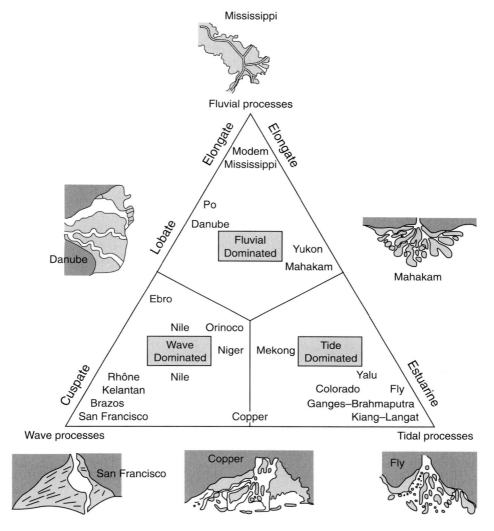

Figure 4.2 Classification of modern deltas based on dominant processes (waves, tides and rivers).
Source: Briggs *et al.* (1997: 311, fig. 17.9).

and 1996. Accretion in areas of the delta subjected to riverine flooding averaged 13.4 ± 7.0 mm/year, whereas areas open to marine conditions accreted only 1.2 ± 0.5 mm/year. The resulting surface elevation in the riverine areas was 11.3 ± 6.1 mm/year, with the difference accounted for by compaction of the accreted sediment. The accretion rate in the marine areas of the delta is less than the current rate of sea-level rise for the region, so that these marine areas, which are so valuable to wildlife, are being submerged. Part of the reason for sediment limitation appears to be extensive engineering works on the delta, that in attempting to prevent flooding have led to the entrenchment of the Rhône and its tributaries and a more effective transport of sediment to the sea (Arnaud-Fassetta, 2003).

4.2.2   Tide-dominated deltas

These include deltas where sediment delivered to the coast by a river is redistributed by tidal currents. They possess tidal channels as well as river channels which creates a rather fragmentary finger-like shoreline. This can be further enhanced by parallel-orientated bedforms, such as tidal shoals, bars and inlets, situated within larger tidal channels (Plate 4.1). The Ganges–Brahmaputra Delta of India and Bangladesh falls within this category.

Plate 4.1 A mangrove-lined tidally influenced channel, with sandy mega-ripples. A distributary channel of the Barron River Delta near Cairns, Queensland (Australia).

Plate 4.2 The linear shore of the wave-dominated Shoalhaven Delta, New South Wales (Australia).

4.2.3   Wave-dominated deltas

These occur where river-delivered sediment is redistributed by wave action. The onshore and alongshore sediment transport promoted by waves tends to produce rather linear or smooth shorelines (Plate 4.2). Where incident waves break parallel to the shore then the delta may develop symmetrically about the river outlet; however, where waves break obliquely to the shore then the delta morphology may be highly asymmetrical because longshore currents transport sediment downdrift. The São Francisco Delta of Brazil is an example of a wave-dominated delta.

The Volta Delta in west Africa is asymmetrical and has recently become predominantly influenced by wave activity that has had far-reaching consequences for the delta and the downdrift coastline (Anthony and Blivi, 1999). The delta has always suffered erosion of its shoreline by wave activity, which has transported this material downdrift along the coasts of Togo and Ghana, maintaining a barrier and lagoon system. Sand lost in this way by wave erosion was always replaced by river sediment derived from upland areas in the river catchment. However, in 1961 the Akosombo Dam was constructed on the River Volta which reduced riverine sediment input into the delta. Since then, erosion by wave activity has dominated the delta shore and the barriers immediately downdrift of the river mouth (see also Box 4.1).

Case Study Box 4.1

The Danube Delta under wave attack

The Danube Delta on the Black Sea coast of eastern Europe is strongly affected by longshore currents. This drift is estimated to transport as much as 2,350,000 m³/year of sediment southward (Giosan *et al.*, 1999). Shoreline changes have been measured for the period 1962 to 1987, indicating that extensive wave erosion is occurring along many stetches of the delta coast. This has been attributed to a reduction in sediment being delivered to the coast by the Danube River, and the construction of shore-protection structures, like groynes, that interrupt the longshore currents leaving local downdrift coastal sections deficient in sediment. A major barrier island is rapidly developing downdrift of the delta as a sink for the transported sediment.

4.2.4   Alternative delta classifications

Of course, within Galloway's (1975) model, two or more of the processes discussed above may be equally important so that some deltas may be river/tide-dominated, tide/wave-dominated or river/wave-dominated, and some may be influenced more or less equally by all three. For example, the Niger Delta, on the west African coast, possesses a concentric pattern of dominant processes, comprising an inner river-dominated zone, an outer wave-dominated zone, and a transitional tide-dominated zone. Similarly, the Red River Delta of Vietnam is influenced by the three processes, but manifested in sections, so that the section of the delta facing the fetch direction is wave-dominated, but the opposite side of the delta is sheltered and is dominated there by mesotidal conditions (Mathers and Zalasiewicz, 1999).

Criticisms of Galloway's (1975) simple river-tide-wave-dominated model have been made in that it takes no account of, for example, sediment grain size and variations in sea

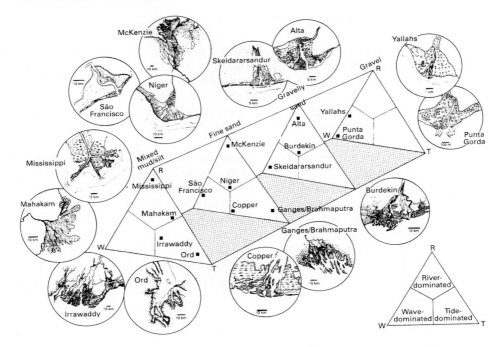

Figure 4.3 Extension of the delta classification scheme outlined in Fig. 4.2 to incorporate sediment particle size.

Source: reproduced with permission from Reading and Collinson (1996: 157, fig. 6.2), and with permission from the Houston Geological Society who are the publishers of Galloway (1975).

level. The significance of grain size has been addressed by Postma (1990), for coarse gravel river-dominated deltas only, and Orton and Reading (1993) who have combined the range of grain sizes and processes to extend Galloway's ternary diagram into a prism (Fig. 4.3). Gravel and mud/silt comprise the two end members, with gravelly sand and fine sand as categories transitional between these end members. The influence of sea-level change on delta development has been reviewed by Boyd *et al.* (1992), who, based on earlier researchers, present a model showing that the regression or transgression of a delta shoreline is due to a combination of the rate of relative sea-level change and the rate of sediment supply (see section 5.3).

4.3 Deltaic sediment supply

By definition, deltas are supplied with sediment from a fluvial source, and globally approximately 15×10^9 tonnes per year of sediment is transported into the sea by rivers. The type of sediment and the amount delivered to the coast is influenced by:

- the size and relief of the river catchment
- the bedrock and its tectonic behaviour
- the local climate, soil and vegetation
- the extent of human activity in the catchment.

Reading and Collinson (1996) suggest that large river catchments are associated with high volumes of fine-grained sediment (with coarser material being stored in the river system),

Scientific Box 4.2

Global significance of the Ganges–Brahmaputra Delta sediment flux

One of the largest deltas in terms of size and sediment flux is the Ganges–Brahmaputra Delta in the Bay of Bengal. Sediment gauges in the river system indicate an annual sediment discharge of 10^9 tonnes/year. The delta has developed since sea level stabilised approximately 7000 years ago, and comprises 1500×10^9 m^3 of sediment in its subaerial part and 1970×10^9 m^3 in its submarine part, which in this assessment includes the deep-sea Bengal Fan (Goodbred and Kuehl, 1999). The significance of this immense sediment load lies in the fact that its source is the vast mountainous regions of the Himalayas and the Tibetan Plateau in central Asia. It is now widely accepted that the weathering of these mountains is consuming atmospheric carbon dioxide, and that river erosion transports this material, carbon and all, to the sea where it is locked away in marine sediments. It is this process that is now thought to be a major influence in the initiation of global cooling and the onset of the Quaternary ice age (Raymo and Ruddiman, 1992).

lower gradients, and high and consistent discharges. Small catchments, on the other hand, are characteristic of tectonically active coasts, which are commonly uplifting with high gradients and energetic streams delivering coarse material to the coast. Climatic relationships are also evident, in that semi-arid regions usually supply sand-grade sediment to the coast, whilst the humid tropics are more often associated with fine-grained silts and clays.

The main reason why the sediment delivered by rivers remains at the coast, enabling deltas to grow and develop, is that shoaling and breaking of waves favour the net landward transport of sediment. This hindrance to offshore sediment movement, with the exception of fine silts and clays, is known as the **littoral energy fence**. This fence may be overcome or bypassed under certain circumstances, along any coast, to allow sediment to move offshore, by:

- **river mouth bypassing** where rivers in flood jet out to sea due to unusually high discharge;
- **estuary mouth bypassing** where strong ebb-tidal currents draw sediment offshore; and
- **shoreface bypassing** where storm wave conditions create strong offshore currents close to the sea-bed.

4.4 River discharge characteristics

The manner in which sediment and water are discharged from a deltaic river into the sea is determined by a complex set of factors, such as the density difference between the fresh and saline waters of the river and sea, respectively, the sediment type and concentration, water depths, water discharge volume, and river velocity (Reading and Collinson, 1996). These in turn influence:

(a) homopycnal flow *(river = sea/lake water density)*

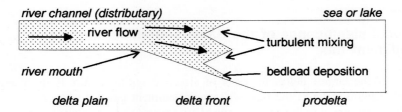

(b) hyperpycnal flow *(river > sea/lake water density)*

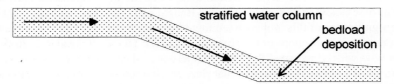

(c) hypopycnal flow *(river < sea/lake water density)*

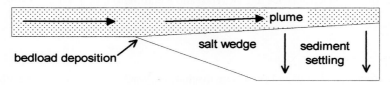

Figure 4.4 Three types of delta hydrodynamics based on the density differences between river and sea/lake water: (a) homopycnal, (b) hyperpycnal, and (c) hypopycnal flow.

- *inertia* of the river flow, that is the ability of the river water to continue moving into the sea before it is slowed or halted;
- *friction* between the flowing water and the bed and sides of the delta river channel, and the sea-bed beyond the river mouth; and
- *buoyancy* of the river water as determined by water density and/or water temperature differences.

Buoyancy is considered highly significant in delta development, and three buoyancy-controlled flow regimes have been recognised: **homopycnal, hyperpycnal** and **hypopycnal flow** (Fig. 4.4).

4.4.1   Homopycnal flow

This occurs where the density of both the river water and the receiving water body are approximately equal. Sediment undergoes turbulent mixing at the river mouth, and most deposition occurs at this location, especially of the bedload. Homopycnal flow typifies deltas that form in fresh water lakes, and is rare in the marine environment.

4.4.2   Hyperpycnal flow

This involves river water that is more dense than the receiving water body, so that the river water flows along the bed. Again, it is characteristic of deltas in fresh water lakes, where the inflowing river is colder than the lake water, and often heavily laden with sediment. Stratification between the cold bottom flow and the warmer upper waters is most pronounced when the river enters deep water. River inflow into shallow water, however, forces a degree of mixing between the two layers which tends to obscure stratification.

4.4.3   Hypopycnal flow

This occurs where the river water is less dense than the receiving water body. This is commonly the case for rivers entering the sea, and hypopycnal flow typifies marine coastal deltas. Here, river water extends as a buoyant jet out into the sea, overlying the denser sea water. Furthermore, upon leaving the river mouth, the river water separates from the bed creating a salt-wedge (similar to that discussed for estuaries in section 3.5.3.1), and also separating bedload from suspended sediment. In this situation, coarse bedload is deposited at the tip of the salt-wedge, where the river water becomes detached from the bed, and fine-grained suspended sediment is transported seaward within the buoyant plume of river water.

4.4.4   River mouth types

The influence of inertia, friction and buoyancy determines the configuration and morphology of deltaic river mouths. Three river mouth types are recognised:

1. **Inertia-dominated river mouths** form where a steep coastal slope occurs seaward of the river mouth, so that the river water is able to expand laterally as well as vertically, building up a uniform lobate delta morphology.
2. **Friction-dominated river mouths** are associated with very shallow water, so that considerable friction between river flow and sea-bed forces the river flow to extend in a lateral direction only. This results in a generally long and narrow sediment build-up along the coast, which is dissected by outflow or tidal channels.
3. **Buoyancy-dominated river mouths** are characterised by hypopycnal flow, so that a relatively straight jet of river water extends out to sea, with little or no lateral extension along the coast. The bedload and the coarse suspended load is deposited at the river mouth as a distinct high bar, whilst the finer sediment is transported further out to sea before being deposited.

Case Study Box 4.3

The buoyant plume of the Rhône Delta, France

The Rhône Delta is located on the Mediterranean coast of southern France. The waters of the River Rhône enter the Mediterranean and extend far offshore as a readily identified buoyant plume. This plume has been mapped using radar techniques and has been studied through physical and geochemical methods (Broche *et al.*, 1998). The results indicate that the morphology of the plume changes in terms of its orientation and in the distance reached offshore. Its orientation is largely determined by surface wind conditions, and the offshore limit largely reflects river discharge. The vertical structure of the plume in the water column also varies, from the classic two-layered situation, with the less dense plume overlying sea water, to a more mixed condition. Also, a complex multi-layered vertical structure has been observed. Studying the dynamics of deltaic fresh water plumes is important because of the exchange of material that takes place from the riverine to the marine system, and therefore is significant in understanding water, sediment and geochemical budgets in the coastal zone.

4.5 River deltas

River deltas include all the major river deltas of the world, where generally large and low-gradient rivers enter the sea (Table 4.2). The low to moderate energy of these rivers supplies mud, silt and sand to the delta where it is reworked to varying degrees by wave and tidal processes (see section 4.2). Terminology used to describe delta sedimentology, morphology and processes is given in Table 4.3. The classic sedimentary structure of deltas (Fig. 4.5), first proposed by Gilbert (1885), but now known not to be universally applicable, comprises:

- subaerial horizontally deposited **topset beds**, laid down by the river and associated channels on the delta platform;
- **foreset beds** that are inclined draping the submerged seaward slope of the delta; and
- **bottomset beds** which are horizontally bedded deposits that occur seaward of the delta slope and are laid down by the settling out of suspended sediment.

Deltas that clearly possess topset, foreset and bottomset beds have become known as Gilbert-type deltas.

Of the terms given in Table 4.3, the process-based terminology is now widely used, and is a useful scheme for discussing the dynamic relationship between deltaic processes and morphology (Fig. 4.6).

4.5.1 Delta plain

The delta plain comprises an extensive lowland environment, dominated by a river channel or a number of **distributary channels** that branch off the main river as it approaches the sea (Plate 4.3). The channels are lined with levées (Plate 4.4), and in between the channels a number of different environments might occur, such as floodplains, lakes, salt or fresh water marshes, mangrove or fresh water swamps, tidal flats, sabkas, and marine embayments. Delta plains have been subdivided into upper and lower subzones (Reading and Collinson, 1996), with the lower delta plain being characterised by the influence of tidal processes and the incursion of salt water, whereas the upper delta plain is dominated by fluvial processes. These subdivisions are perhaps not well named, as there may not be any significant altitudinal differences between the two subzones. Since their recognition is determined by the degree of marine inundation into the delta, terms such as outer and inner delta plain may be appropriate substitutes for lower and upper subzones, respectively, and are used hereafter.

Table 4.2 Examples of deltas and their river, sediment, delta plain and coastal environment characteristics (Reading and Collinson, 1996)

	River characteristics			Sediment		Delta plain		Coastal environment		
delta	catchment (10³ km²)	discharge (m³/s)	sediment load (10⁶ tonnes/yr)	sediment: water ratio (g/l)	grain size (mm)	area (km²)	gradient (m/km)	wave height (m)	tidal range (m)	water depth (m)
Amazon	6150	199,634	900	0.14	0.03	467,078	0.0125	moderate	4.9	100
Ebro	85.8	552	6.2	0.35	0.2	325	0.38		0.2	100
Fraser	234	3549	20	0.18	0.12–0.35	480			5	350
Ganges–Brahmaputra	1597.2	30,769	1670	1.76	0.16	105,641	0.05–0.17	low	3.6	
McKenzie	1448	9100	126		silty sand	13,000	0.05	low	0.2	70
Mekong	790	14,168	160	0.37	fine sand	93,781	0.02	low	2.6	
Mississippi	3344	15,631	349	0.73	0.014	28,568	0.02	very low	0.4	
Niger	1112.7	8769	40	0.08	0.15	19,135		moderate	1.4	
100–200										
Nile	2715.6	1480	111	2.43	0.03	12,512	0.088	0.5–1.5	0.4	100
Orinoco		34,856	210	0.19		20,642	0.067	low	1.9	
Po	71.7	1484	15	0.33	0.52	13,398	0.025–0.074		0.7	
Rhône	90	1552	10	0.21	0.08–0.5	2540				50–100
São Francisco	602.3	3420	6			734		1	2.5	
Senegal	196.4	867.8				4254		2.5	1.9	
Shoalhaven	7.25	57			0.25	85		1.5	1.2	
Yangtze	1354.4	28,519	478	0.54	silt	66,669		1–1.5	2.8	50

Table 4.3 Comparison of terms describing the sediment structure, morphology and processes that operate in the three major deltaic areas (see also Fig. 4.4).

Sedimentological (Gilbert, 1885)	Morphological (Nittrouer et al., 1986)	Process (Reading and Collinson, 1996)
topset beds	delta platform	delta plain
foreset beds	delta slope	delta front
bottomset beds	prodelta	prodelta

Source: after Park (1997).

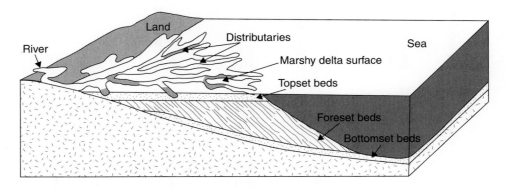

Figure 4.5 General structure of a delta, indicating the location of topset, foreset and bottomset beds.

Source: Park (1997: 325, fig. 11.25).

On a broad scale then, delta plains comprise distributary channels and the areas between the channels, known as **interdistributary areas**. Channels on the inner delta plain behave similarly to fluvial channels. They normally have unidirectional flow and may meander according to the gradient. Channels on the outer delta plain, however, behave more like estuaries (see section 3.5), in that flood tides penetrate the river mouths and may pond the river water, reducing water and sediment discharge, and in some cases where river mouth mixing is minimal, a distinct salt-wedge may occur. Both fairweather and storm wave activity may also affect inner delta plain channels, which again may lead to ponding back of the river water. River ponding like this may result in flooding, and in some cases the switching of channels, from the old pre-flood channel to a new post-flood channel.

In addition to the riverine distributary channels, there may also be so-called tidally influenced channels that occur as shallow embayments or tidal inlets penetrating the outer edge of the delta plain. These are unconnected to the river system and may often be distinguished by an overall funnel-shaped form, narrowing inland, whereas river channels are usually parallel-sided.

The interdistributary areas can support a range of environments. Where they are vegetated a zonation may be apparent, with the most landward areas being characterised by fresh water swamps that become brackish marshes or mangroves as the sea is approached. Areas between tidally influenced channels are usually intertidal, comprising salt marshes, mangroves or sabkas depending on the climatic setting, and in some cases dense tidal creek networks exist. Alternatively, interdistributary areas may be permanently or semi-permanently flooded to form shallow lakes or lagoons. A number of geomorphological features may occur in interdistributary areas, such as:

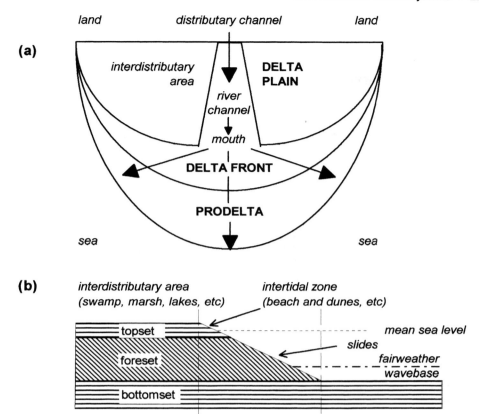

Figure 4.6 Schematic geomorphology of a river delta: (a) plan view; (b) shore-normal cross-section.

Plate 4.3 The extensive delta plain of the Shoalhaven Delta, New South Wales (Australia).

Plate 4.4 Cows grazing on levées of the Shoalhaven River, part of the Shoalhaven Delta, New South Wales (Australia).

- abandoned channels and levées;
- **crevasse channels and splays**, which are formed where levées have been locally breached to produce a crevasse channel, and from which a crevasse splay may extend for short distances into the interdistributary area;
- minor deltas, which may grow from an initially small crevasse splay and develop to such an extent as to infill a low-lying interdistributary area.

4.5.2 Delta front

The delta front zone includes the river mouth, the shoreline of the delta, and the area immediately offshore. The geomorphological features encountered in this zone are very much dependent on the processes discussed in sections 4.2 and 4.4. For example:

- River-dominated deltas tend to create river mouth sand bars that are orientated perpendicular to the river flow, and also levées that may descend below sea level, defining a submarine channel.
- Tide-dominated deltas produce tidal channels and inlets separated by sand bars, that are usually aligned parallel to the tidal flow. Flood or ebb-tidal deltas may also be created within the river mouth.
- Wave-dominated deltas are characterised by beaches, spits and longshore bars, where wave activity has moved sediment onshore. Cheniers and aeolian dunes may also be present.

Of course, dominant delta processes may change through time and under different conditions. For example, under normal discharge conditions a river mouth may be buoyancy-dominated with a well-defined intruding salt-wedge. However, with higher discharges, associated with storm rainfall for instance, the river mouth may become inertia- or friction-dominated. These changes will be reflected in the morphological features present at a given time (Reading and Collinson, 1996).

4.5.3 Prodelta

This occurs in the form of an apron around the foot of the sloping delta front, lying in relatively deep water and characterised by fine sediment. The prodelta is below wave-base and so not affected by waves, and tidal influences are also minimal. Where the delta is in shallow water and the foot of the delta front is above wave-base, a fine-grained prodelta is unlikely to form. Prodelta sediment is deposited out of suspension from river water, outflowing into the sea as a buoyant plume. Laminations in prodelta sediments are common and represent regular fluctuations in the grain size of sediment discharge from the river at times of different energy regimes, for example summer low-energy discharge and winter high-energy discharge. These laminations are best preserved where the prodelta bottom waters are anoxic, otherwise the burrowing activity of benthonic organisms may destroy these subtle sedimentary textures (Reading and Collinson, 1996).

4.6 Fan deltas

This is a collective term for a number of different delta types that are characteristically composed of coarse-grained sediment, rather than the finer material of a river delta. The coarseness of the sediment is indicative of the relatively high fluvial energy involved in transporting the material to the coast. For **fan deltas**, the fluvial source is known as the **feeder system**, and Postma (1990) has recognised three distinct feeder systems:

1. A type A feeder system is a single high-gradient river or stream channel, that runs off high-relief areas, often tectonically uplifting. Local landslides as well as stream flood events deliver very coarse gravel sediment as bedload to the delta front and sometimes to the prodelta region.
2. A type B feeder system is characterised by a relatively high gradient, but this time a **braided river** (a single river course that comprises multiple channels) is involved. This type also occurs on tectonically active coasts, but is especially common in glacial environments, where the braided rivers are fed by glacial meltwater. In such cases, cold meltwater discharge may be very dense and the resulting hyperpycnal flow commonly transports coarse gravel material down the delta front to the prodelta.
3. A type C feeder system is characteristic of a moderate-gradient braided river or **braidplain** (a plain with several braided rivers), and often occurs in glacial environments. Despite the lower energy, however, sediment is still predominantly transported as bedload, but the sand content is usually higher than coarse gravels. Based on sedimentology, this feeder system type may be thought of as transitional between the true high-gradient fan deltas and the low-gradient river deltas discussed in previous section 4.5 (Reading and Collinson, 1996).

4.6.1 Fan delta morphology

Fan deltas are very varied in their morphology as they often occur in tectonically active coasts, which may over time vary river gradients, and therefore energy inputs, and relative sea level. However, three broad types have been recognised (Fig. 4.7) on the basis of gradient and water depth of the coastal shelf (Reading and Collinson, 1996):

1. *Shelf-type (shallow water) fan deltas* occur on low-gradient and shallow-water coastal shelves. The overall morphology is that of a gradually seaward sloping delta plain

and front, extending gently down to the prodelta. The sediment is generally coarse, but a seaward decrease in grain size is often seen, and the prodelta (if the water is not too shallow) is composed of relatively fine-grained sediment. This represents a lateral zonation of sediment types.

2. *Slope-type (deep water) fan deltas* occur in deep water and are characterised by a steep slope between the delta front and prodelta. In certain cases, there may be a submarine delta at the base of the steep slope, fed by submarine flows down the delta front. Generally, the overall morphology is characterised by a low- to moderate-gradient delta plain and a high-gradient delta front. Lateral sediment zones include a gravelly delta plain that passes seaward into extensive beach deposits, that are perched at the top of the steep delta front.

3. *Gilbert-type fan deltas* are so-called because they possess the distinctive topset, foreset and bottomset sediment layers that were first described by Gilbert (1885) (see section 4.5). These fan deltas are steeply inclined throughout, from delta plain to prodelta, and may occur in both deep and shallow water. Mass movement proceses are commonly active on these steep deltas, where slope gradients may be as much as $35°$.

(a) shelf-type (shallow water) fan delta

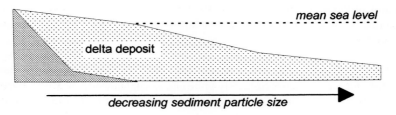

(b) slope-type (deep water) fan delta

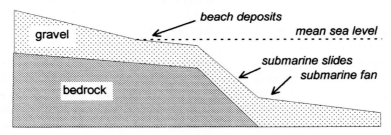

(c) Gilbert-type fan delta

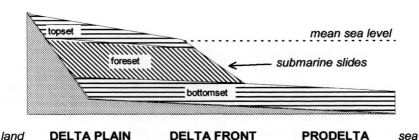

| land | **DELTA PLAIN** | **DELTA FRONT** | **PRODELTA** | sea |

Figure 4.7 Schematic geomorphology of fan deltas: (a) shelf-type (shallow water) fan delta; (b) slope-type (deep water) fan delta; and (c) Gilbert-type fan delta.

Because of the coarse-grained nature of fan deltas, tides have little influence on sediment transport and reworking. Similarly, fairweather wave conditions play a limited role in this respect, but storm wave events and tsunami may be significant, particularly considering the high latitude and tectonically active locations where fan deltas are often encountered.

4.7 Holocene delta development

Sea levels during the Quaternary ice ages were much lower than today. Indeed, during the Last Glacial Maximum at 18,000 years ago, sea levels were approximately 130 m lower than at present, and the world's shorelines were located close to the edge of the continental shelf. During the post-glacial Holocene Epoch, sea levels have risen, rapidly at first to about 6000 years ago, and then more slowly. Deltas that were active during the Last Glacial Maximum and formed at the continental shelf edge were either drowned by Holocene sea-level rise, or survived if the rate of sediment supply, accretion and elevation of the delta surface always exceeded the rate of sea-level rise. Where these deltas do survive they are termed **shelf-edge deltas** (also known as shelf-margin deltas), and they are often surrounded by relatively deep water (Porebski and Steel, 2003). Where deltas did not elevate faster than sea level, then they migrated landward, always occurring in relatively shallow water and known as **inner-shelf deltas** (also known as shoal-water deltas). Most of the largest modern deltas, that protrude out onto continental shelves, are of this type (Reading and Collinson, 1996).

Vertical sediment sequences in many modern deltas reflect the onshore migration of delta environments. A typical Holocene sedimentary sequence may comprise a basal layer of riverine sediments, such as floodplain deposits, overlain by delta plain distributary channel and interdistributary area deposits, passing upwards into tidal sediments, and finally shoreline deposits, which may include beach-dune sands (e.g. Warne *et al.*, 2002). This sequence simply reflects the upward and landward migration of the various delta environments.

Such sedimentary responses to rising sea levels during the Holocene are experienced throughout the deltaic river system. Aslan and Autin (1999) report sedimentological changes through a 15–30 m thick Holocene sequence at a location 300 km upstream from the mouth of the Mississippi Delta. Here they find two sediment units:

1. a Lower Holocene unit, deposited prior to 5000 years ago, consisting of swamp and shallow lake muds, and deposits from numerous small stream channels; significantly there is little evidence for soil formation in the river environment at this time; and
2. an Upper Holocene unit, deposited from 5000 years onwards, consisting of sand bodies representing river meanders, silts from levée formation, and evidence for soil formation.

Therefore, a dramatic change occurred 5000 years ago in the depositional environment, which has been interpreted as representing the deceleration of the rate of Holocene sea-level rise. During rapid rise the Mississippi floodplain accreted rapidly, but slowed when sea levels stabilised 6000 years ago, which increased river meandering and allowed soils to form. Clearly, this location, presently 300 km inland, is affected by coastal processes and therefore indicates how wide the coastal zone is if we strictly adhere to the definition introduced in section 1.1.1. Törnqvist *et al.* (2004) also consider the Mississippi Delta to be subsiding in response to the melting of the Laurentide ice sheet at the end of the last glacial.

Scientific Box 4.4

Delta response to sea-level stabilisation

Following the onshore migration of a delta and the stabilisation of Holocene sea level, a delta may prograde seaward if sediment supply is great enough. An example of this is the Shoalhaven Delta in New South Wales, Australia (Young *et al.*, 1996; Umitsu *et al.*, 2001). The Shoalhaven River is a major drainage system on the southeast Australian coast, with a substantial fluvial sediment load (Fig. 4.8; Table 4.2). It appears that upon sea-level stabilisation 6000 years ago, the Shoalhaven River extended across a large embayment, as a levée-lined channel, to a wave-dominated sand barrier to which it supplied sediment; its descendant is now known as Seven Mile Beach. The deltaic back barrier area subsequently infilled with fluvial sediment, resulting in the seaward progradation of the shoreline by 3.5 km. The delta continued to be influenced by tides up until 2500 years ago, but this influence declined as fluvial infilling progressed, and although the mouth of the Shoalhaven River is still tidal, tidal processes are no longer significant for much of the delta. Therefore, for many deltas rapid sea-level rise results in landward retreat, whereas sea-level stabilisation in conjunction with high sediment input leads to seaward delta progradation (see Plates 4.2–4.4).

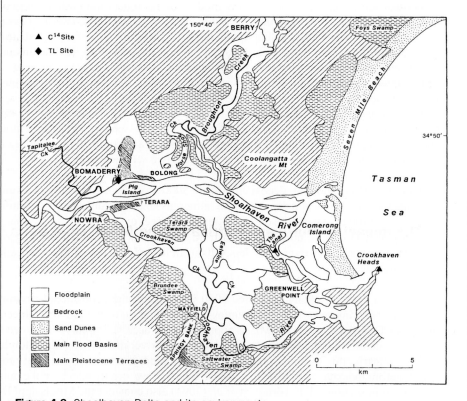

Figure 4.8 Shoalhaven Delta and its environments.

Source: reproduced with permission from Young *et al.* (1996).

4.8 Deltas and human activity

Deltaic areas have long been attractive to human populations, due to a variety of reasons.

- *Topography* – deltas are usually of a uniform low relief, which makes communications routes within deltas relatively easy to establish, with the use of ferry boats, fords and bridges to cross waterways.
- *Fertility* – flooding of the delta plain supplies silt to the soil which renews its fertility. Therefore, deltas are excellent agricultural locations, and many countries have relied on harvests from deltaic areas for subsistence. This includes the Nile Delta which supported the Egyptian dynasties for thousands of years.
- *Reclamation* – new, highly fertile agricultural land can be brought into service through the construction of modest drainage systems in the wettest parts of the delta plain and front (e.g. swamps, marshes, shallow lakes), so increasing productive land.
- *Trade* – the rivers that create deltas are often large enough to be navigable to ocean-going craft, and therefore major ports often arise in deltaic settings to import goods from abroad, and to export goods, including agricultural produce derived from the delta itself.

However, some of these attraction factors can also present problems. This is especially true of the floods involved in depositing silts. Bandyopadhyay (1997) lists a considerable number of natural hazards that are affecting Sagar Island (India) in the western part of the Ganges–Brahmaputra Delta. The island was reclaimed from mangrove wetlands from 1811 onwards, and is now fully populated. The effects of tropical cyclones, coastal erosion, tidal inundation and sand dune encroachment are just some of the hazards faced by the islanders. Addressing these hazards has led to seven different agencies having some involvement in protecting the island, often with little coordination. Protection schemes that have been initiated include the building of embankments, both along river courses and at the shoreline, replanting of mangroves and vegetation wind-breaks, resettlement areas for displaced people, and refuges for when storms occur. The success of these schemes is variable.

Case Study Box 4.5

The impact of Hurricane Katrina on the Mississippi Delta in 2005

Hurricane Katrina was the third most severe hurricane to make landfall on the Gulf Coast of the United States. News reports state it came ashore on the Mississippi Delta, a classic bird's-foot delta protruding out into the Gulf of Mexico, on 28 and 29 August 2005. The Category 5 hurricane had sustained winds that peaked at 280 km per hour, generated a storm surge 9 m high, and spawned storm waves up to 17 m high. In the city of New Orleans the delta levées were breached by the storm surge in over 50 places, which led to widespread flooding up to 4.5 m deep locally, and penetrating to over 16 km inland. The human cost was great, with 60,000 residents made homeless and 2500 people confirmed to have died or to be missing after the disaster. Many commentators suggest the economic cost of Hurricane

Katrina is in excess of 100 billion US dollars with initial estimates as high as 150 billion US dollars (Burton and Hicks, 2005).

The coastal impacts of the hurricane were varied, including coastal erosion and migration of barrier islands fringing the Mississippi Delta (see Plate 4.5). However, one of the most interesting physical impacts of Hurricane Katrina, and Hurricane Rita that occurred later that year, was the delivery and deposition of sediment to the surface of the coastal wetlands. Turner *et al.* (2006) estimate that 412 billion tonnes

Plate 4.5 Before and after views showing the impact of Hurricane Katrina at Biloxi along the Gulf of Mexico coast. Note the destruction of a number of large buildings and other structures, the large ripple bedforms on the beach (bottom left of lower photograph), and sand transported inland.

Source: Reproduced with permission of the United States Geological Survey.

of sediment was deposited on the surface of the wetlands, and in the open water areas, by hurricane-linked flooding in 2005. This vast amount of sediment is equivalent to 12% of the annual suspended load of the Mississippi River and is 5.5 times greater than the amount that was naturally deposited by seasonal river flooding prior to the building of levées for flood protection. Therefore, hurricane-linked flooding appears to be a major factor in maintaining healthy levels of sedimentation in these coastal wetlands. This is important because there are significant concerns that the Mississippi Delta ground surface is subsiding, due to sediment compaction, faster than it is accreting (Törnqvist *et al.*, 2006). The deterioration of the Mississippi Delta plain due to human activities, such as protection measures to stop river flooding, probably contributed to the severity of the disaster, and efforts are now being made to re-establish natural system links between the river and delta plain (Day *et al.*, 2007).

In another part of the Mississippi Delta system, on the continental shelf, sediments also appear to accumulate preferentially during hurricane events with sediments liberated by coastal and shallow-water erosion being transported out to sea. Indeed, Dail *et al.* (2007) suggest that up to 75% of shelf deposition occurs during severe hurricane events. The submarine Mississippi Canyon, located offshore, is one such high-deposition site that subsequently acts as a sediment sink, preventing sediment from re-entering the coastal system. It is clear then that Hurricane Katrina, like other severe hurricanes, not only had a high human and economic cost, but also imposed a major reorganisation of coastal landforms and sediments within the Mississippi Delta system.

4.8.1 Human impacts on deltas

Many of the human impacts already discussed Chapters 2 and 3 apply equally to deltaic wave and tidal environments. However, impacts that are specific to deltas have come to light since the 1970s, and originate from interference with the river systems that supply deltas with water and sediment (Fig. 4.9). Interference includes the following:

- *The expansion and intensification of agriculture within a river catchment*, including deforestation, may increase soil erosion and in turn the sediment supply to the delta, via the river system. This may lead to the silting up of distributary channels, which could increase the frequency and magnitude of delta plain flooding. Also, if the sediment is discharged from the river mouth, it may enhance shoreline sedimentation and the seaward progradation of the delta front.
- *The construction of dams in the river system* is known to severely affect delta dynamics. Both water and sediment discharges are reduced, which limits the capability of a river to flush itself out of sediment and pollution, and reduces the sediment supply to the delta front, which often leads to shoreline erosion.

The latter impact is of concern in many major deltas worldwide, as many cities built on delta fronts are threatened by coastal erosion and inundation.

The Colorado River has had its flow controlled to the point of elimination for nearly a century, and the future of its delta is being determined by wave and tidal conditions in the Gulf of California. The reduction in fresh water discharge into the upper Gulf of California has forced a change from brackish water to hypersaline conditions, and sediment analysis

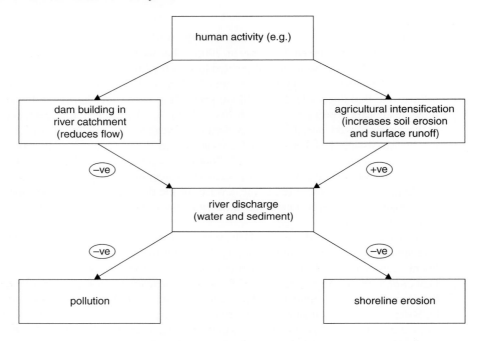

Figure 4.9 The relationships and feedbacks between some examples of human activity and the dynamics of a delta system.

shows that Colorado Delta sediments are currently being extensively reworked (Carriquiry and Sanchez, 1999). Similarly, the water and sediment supply to the Luanhe River Delta in northern China has decreased rapidly since 1980, following water conservancy projects constructed in the river catchment. The Luanhe Delta only started to develop in its present position in 1915, due to the river switching its channel, and human interference is already leading to its abandonment (Feng and Zhang, 1998).

Sediments along the shoreline of the Nile Delta in Egypt are being redistributed by wave activity, as a consequence of the construction of the Aswan Dam in 1964, and could have serious physical and socio-economic impacts, especially if sea levels rise in the near future (El Raey *et al.*, 1999a, b). Such a future sea-level rise (see section 5.4 for details) is predicted to have major impacts on human populations on many deltas. The Yangtze Delta in China is a peri-urban area around Shanghai, with significant agricultural activity. Chen and Zong (1999) suggest that future sea-level rise will result in erosion along the southern coast of the delta, and that sediment infilling in parts of the Yangtze Estuary will occur. It is argued that the risk of flooding will increase, that the water table will rise and induce prolonged waterlogging, and that during the dry season tidal incursions of sea water will reduce the amount of fresh water available for the irrigation of agricultural lands, which may lead to **salinisation**. Altogether, it is likely that the agricultural productivity of the region will be greatly reduced.

These examples show how sensitive delta systems are to human interference, and how extensive the system feedbacks are, so that actions taken far inland in the river catchment affect the environment and people living at the coast. Because of the magnitude of interference, deltas are the focus for many international monitoring studies; however, because of the spatial scale involved, remote sensing techniques are widely employed, and are perhaps the only way of effectively recording erosion, accretion and general delta shoreline changes through time (e.g. White and El Asmar, 1999).

Summary points

- Deltas differ from other coastal environments in that their principal sediment source is derived from river systems, yet they are still influenced strongly by wave and tidal activity.
- Delta hydrodynamics are determined by the density differences between the river and sea water, which in turn largely determines the fate of river-borne sediment within the delta system.
- Delta geomorphology comprises a subaerial delta plain, a delta front and a submerged prodelta.
- The development of modern deltas has been strongly influenced by the post-glacial rise in sea level, which has caused deltas to retreat landward before prograding seaward after sea levels stabilised.
- Human impacts on delta systems are varied. Of special concern is interference with delta river-system catchments, by agricultural intensification and dam-building projects for example, which influence water and sediment discharges at the river mouth, often with feedbacks that alter shoreline positions.

Discussion questions

1. Assess the validity of using the contribution of wave, tide and river activity to classify deltas.
2. Examine the links between the hydrodynamics, sedimentology and geomorphology of delta systems.
3. Citing examples, evaluate the physical changes brought about through human interference in the river catchments of delta systems.

Further reading

See also

Tectonic classification of coasts, section 1.3.2
Longshore beach features on wave-dominated coasts, section 2.6.3
The role of tidal currents, section 3.4
Global warming and the threat of future sea-level rise, section 5.4

Introductory reading

Deltaic and estuarine environments. P. French. 2007. In: C. Perry and K. Taylor (eds) *Environmental Sedimentology*. Blackwell, Oxford, 223–262.
A wide-ranging yet brief overview of deltaic sedimentary environments.

Clastic coasts. H. G. Reading and J. D. Collinson. 1996. *In:* H. G. Reading (ed.) *Sedimentary Environments: Processes, Facies and Stratigraphy*. Blackwell, Oxford, 154–231.
A substantial chapter on modern and ancient clastic coasts, but with a clear emphasis on deltaic sedimentary environments.

Coarse-grained Deltas. A. Colella and D. B. Prior (eds). 1990. Blackwell Scientific Publications, Oxford, 357pp.
A compilation of research papers on fan deltas.

Advanced reading

Deltas: Sites and Traps for Fossil Fuels. M. K. G. Whateley and K. T. Pickering (eds). 1989. Geological Society, London, Special Publication No. 41, 360pp.
As the title suggests, this is a collection of geological research papers devoted to hydrocarbon exploration in deltaic regions.

Dating modern deltas: progress, problems, and prognostics. J. D. Stanley. 2001. *Annual Review of Earth and Planetary Sciences*, **29**, 257–294.
An examination of dating methods and associated problems for modern deltas.

Journal of Coastal Research, volume **14**, part 3. 1998.
A thematic issue devoted to deltaic coastal environments, with some excellent review articles, including the Mississippi Delta and Arctic deltas, as well as research papers.

5 Sea level and the changing land-sea interface

Sea-level change is one of the main factors in stimulating coastal change, and in the long term it has been controlled by changes in the volume of landlocked ice sheets. If sea level rises, then coasts must be able to change dynamically to keep up, or drown. On the scale of a human lifetime we may consider sea level to be unchanging, but even through the twentieth and early twenty-first centuries it appears that sea level has been rising due to human activity, and may continue to rise well into the future. This chapter covers:

- **the principal mechanism for sea-level change**
- **methods employed in constructing records of sea-level change**
- **the various responses of coastlines to sea-level change**
- **twentieth-century sea-level rise, that has been attributed to our enhancement of the Greenhouse Effect, and future predictions**
- **ways of managing sea-level rise**

5.1 Introduction

The level of the sea is not constant, it is always rising and falling, whether through the passing of waves (including tides), meteorological influences, or gravitational effects in the form of the earth's **geoid**. Long-term and significant sea-level changes, however, reflect changing levels of both land and sea; **eustasy** refers to absolute changes in global sea level, and **isostasy** refers to the vertical movement of land due to local geological factors. It is the balance between these two processes at a given coastline that produces observed changes in sea level, referred to as **relative sea-level change**, because an absolute (eustatic) rise in sea level may not be required to allow the sea to rise relative to the land. Figure 5.1 explores this balance in a simplified way, and a number of cases can be put forward to illustrate the relationship:

- Relative sea-level rise will occur if: (a) eustatic sea level rises whilst the land is isostatically subsiding, static or uplifting at a slower rate than eustatic rise; (b) eustatic sea level is static whilst the land is isostatically subsiding; and (c) eustatic sea level falls whilst the land is isostatically subsiding at a faster rate than eustatic fall.
- Relative sea-level fall will occur if: (a) eustatic sea level falls whilst the land is uplifting, static or subsiding at a slower rate than eustatic fall; (b) eustatic sea level is static whilst the land is isostatically uplifting; and (c) eustatic sea level rises whilst the land is isostatically uplifting at a faster rate than eustatic rise.
- A sea-level **stillstand** will occur if: (a) eustatic sea-level fall and isostatic subsidence of the land occur at the same rate; (b) eustatic sea level and the isostatic land level are both static; and (c) eustatic sea-level rise and isostatic uplift of the land occur at the same rate.

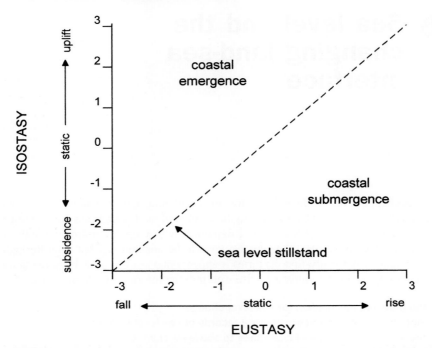

Figure 5.1 The relationship between eustatic and isostatic changes and their influence on relative sea level and coastal emergence/submergence (the numbers on the axes represent rates of change).

The main reason for eustatic and isostatic changes in the level of the sea and land is the growth and decay of continental ice sheets during the Quaternary. Glacierisation requires ice, which is introduced into glacial systems as snow, mainly derived from oceanic evaporation. Therefore, during Quaternary glacial stages water is extracted from the oceans and transferred to ice sheets where it is locked up for the duration of the glacial stage. This glacial control on eustasy is known as **glacio-eustasy**. Similarly, these same continental ice sheets growing on land exert great overburden pressure, causing the land to isostatically subside, a process known as **glacio-isostasy** (Fig. 5.2). At the end of a glacial stage, melting ice returns water to the ocean causing sea levels to rise, and at the same time the release of ice overburden results in the isostatic rebound or uplift of the land. Therefore, previously glaciated areas often possess very complex relative sea-level histories, reflecting the balance between glacio-eustatic and glacio-isostatic changes (e.g. Shennan *et al.*, 2006; Simms *et al.*, 2007; Smith *et al.*, 2007). Furthermore, knowledge of these processes has led to the development of models that attempt to predict evolving coastlines since the last glacial through the construction of shoreline maps for different time periods (e.g. Lambeck, 1995).

The Quaternary Period spans from 1.81 million years ago to the present day and comprises the **Pleistocene Epoch** (1.81 million years to 10,000 years ago) and the **Holocene Epoch** (10,000 years ago to the present). During the Quaternary sea levels have fluctuated widely, with sea-level highstands during the interglacials and lowstands during the glacial stages. Many interglacial sea-level highstands attained or exceeded the present sea-level altitude, and many fossil shorelines may be encountered above present sea level and appear to be 'raised', hence the term **'raised' beach** (Plate 5.1). The word 'raised' can be a misnomer, because many of these fossil Pleistocene shorelines are *in situ*.

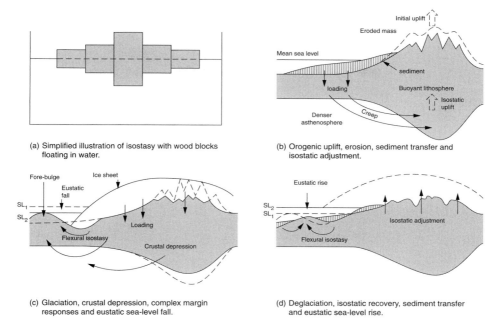

(a) Simplified illustration of isostasy with wood blocks floating in water.

(b) Orogenic uplift, erosion, sediment transfer and isostatic adjustment.

(c) Glaciation, crustal depression, complex margin responses and eustatic sea-level fall.

(d) Deglaciation, isostatic recovery, sediment transfer and eustatic sea-level rise.

Figure 5.2 Stages in glacio-isostasy, from ice-loading and crustal depression to deglaciation and isostatic rebound (SL = sea level).

Source: Briggs *et al.* (1997: 54, fig. 4.7).

In the British Isles, mapping isostatic readjustments following the last glacial at the end of the Pleistocene shows that northern Britain is currently uplifting by up to 2 mm/year, whilst southern Britain is subsiding by up to 2 mm/year. These are reflected in the rates of relative sea-level change (Fig. 5.3). The two regions are separated by a fulcrum line that extends in a stable zone from the Tees Estuary in northeast England, through the Dee Estuary, and on to the Lleyn Peninsula in northwest Wales. This post-glacial isostatic movement is reflected in the distribution of **submerged forest** sites in Great Britain (Fig. 5.4). The majority of sites are located in southern Britain, whilst only a few are found north of the fulcrum line. This indicates that post-glacial eustatic rise coupled with isostatic subsidence in southern Britain has led to relative sea-level rise and the drowning of previously wooded landscapes, whilst in northern Britain isostatic rebound is outpacing eustatic rise, so that relative sea levels are actually falling there and resulting in coastal emergence rather than submergence (Fig. 5.3).

Only in isostatically stable areas of the world can records of eustatic or absolute sea-level change be possibily found, i.e. areas that have never been glaciated, or that suffer earthquakes, volcanic or other tectonic activity (earth movements) that may vertically displace the land. Such areas are rare, but the island of Barbados is considered to be a stable carbonate platform from which a eustatic sea-level record has been constructed and examined (Peltier, 2002). Eustatic sea-level curves from such areas indicate that the last glacial sea level at 18,000 years ago was approximately 120 m below the present sea level. A period of very rapid rise ensued until approximately 6000 years ago, but since then sea level has been comparatively stable with a rise of less than 10 m to the present day. Figure 5.5 shows sea-level rise from 18,000 years ago to the present, and how the coastline of Great Britain has developed.

Plate 5.1
A fossil Pleistocene 'raised beach' at Plage de Mezpeurleuch, Brittany (France): (a) a stratified sequence containing a number of different facies (I-VI) representing different depositional palaeoenvironments; (b) a poorly sorted unit comprising cobbles, pebbles and smaller gravel clasts infilling around large boulders.

Source: Haslett and Curr (2001).

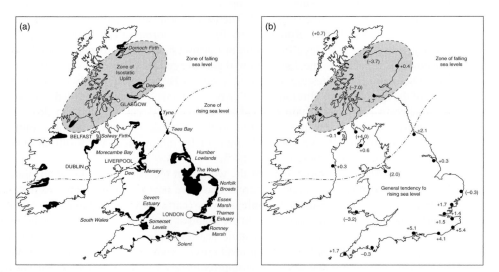

Figure 5.3 Isostatic map of Great Britain showing (a) areas of isostatic uplift (grey shading) and those threatened by sea-level rise (black shading), and (b) rates of sea-level change in (mm/year).

Source: Briggs *et al.* (1997: 320, fig. 17.20).

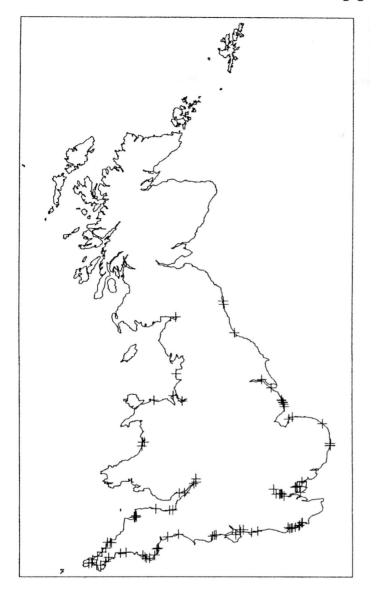

Figure 5.4
The distribution of submerged forests in Great Britain indicating main areas of isostatic subsidence.

5.2 Constructing records of sea-level change

Mean sea level is known to have varied altitudinally through time. Ample evidence exists to suggest that major changes in sea level have occurred through geological time. Evidence occurs in the following forms:

- *Erosional landforms*, such as cliffs, shore platforms, marine notches, sea caves and arches, may now occur well above present sea level, or indeed may now be submerged.
- *Depositional features*, such as fossil beaches, tidal flats and coral reefs, again may now occur either above or below present sea level.
- *Biological indicators*, i.e. organisms which when alive live at known tidal levels, may become stranded by changes in sea level, so that their occurrence as fossils

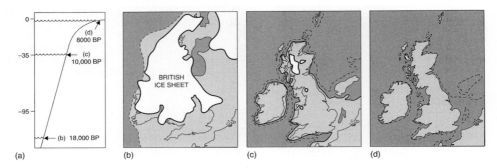

Figure 5.5 Sea-level rise and the development of the coastline of the British Isles. (a) Sea-level rise since 18,000 years ago. (b) Extent of the last ice sheet. (c) Early Holocene coastline. (d) Late Holocene coastline (broken line indicates position of recently abandoned shoreline).

Source: Briggs *et al.* (1997: 308, fig. 1).

indicates the position of the old shoreline; examples include barnacles, tubeworms and oysters.

- *Archaeological remains* can sometimes indicate sea-level change, such as submerged dwellings or ports that have been elevated above sea level. Greece is well-endowed with such remains, but submerged megaliths (standing stones) in Brittany (France) are also excellent examples.
- *Historical documents and records* can indicate recent changes; this includes tide-gauge records that are available for in excess of 100 years at some locations, such as Brest in northwest France.

Changes in sea level through time are most frequently depicted as linear curves, illustrating the altitude of sea level for particular times in the past. These curves should be constructed using **sea-level index points** (SLIPs) which can be any of the sea-level indicators discussed in section 5.1 for which age, altitude, **tendency of sea level** and **indicative meaning and range** are known (see below for explanations). In the past, sea-level curves have been produced with reference to age and altitude only, but these curves indicate very general sea-level trends only and are difficult to compare from region to region, so care should be taken when consulting early sea-level curves.

5.2.1 Dating sea-level index points

Classical archaeological remains and historical documents quite often have good dates assigned to them, based on the period from which they originate, but for natural sea-level indicators, laboratory and field dating methods must be employed. The technique chosen for dating a sea-level index point depends to a great extent on the sea-level indicator under investigation. Erosional landforms are notoriously difficult to date in absolute terms, that is to assign an age in years. However, Haslett and Curr (1998) have attempted, for example, to date the relative ages (i.e. indicating which one is older than another without establishing actual dates) of a number of Quaternary erosional shore platforms in Brittany (France) using the weathering characteristics of the local granite bedrock. This works on the assumption that the oldest platform should be more weathered than platforms created by subsequent sea levels, because they have had more time to weather. Although the results are not straightforward, the technique demonstrates its potential as a useful discriminator of relative age. Erosional features are more easily dated if they have biological remains attached to them, such as fossil barnacles still attached to abraded surfaces, as discussed below.

For Late Quaternary sea-level changes by far the most frequently used technique is the **radiocarbon dating** method. Radiocarbon (^{14}C) naturally occurs in the atmosphere and is incorporated into all living organisms at the same concentration as occurs in the atmosphere. Upon death ^{14}C undergoes radioactive decay, and radiocarbon dating uses the decay of this carbon isotope to date carbonaceous material, including preservable material such as wood, bone, peat and shell. The half-life of ^{14}C was calculated in the 1940s as being 5568 years, and therefore by measuring how much ^{14}C is left in a sample it is possible to estimate time elapsed since death. It is now known that the half-life of ^{14}C is in fact 5730 years, but the initial estimate is still used so that comparison with early radiocarbon dates can be made. The measurement of ^{14}C in a sample is achieved either by the conventional technique of counting radioactive particles emitted from a sample, or by directly measuring the ^{14}C atoms in a sample using accelerator mass spectrometry (AMS) (Kaplan, 2003; Muzikar *et al.*, 2003; Skipperud and Oughton, 2004). Because the AMS technique makes a direct measurement, rather than relying on radioactive emissions, it is more accurate and can be performed on very small samples.

Although radiocarbon dating has become a widely used technique, it is not without uncertainty. Importantly, tree-ring studies (dendrochronology) have shown that radiocarbon years are not the same as calendar years. Beyond 2000 years ago it appears that radiocarbon dates yield consistently younger ages than they should; for example, a radiocarbon age of 6000 years is in fact closer to 7000 calendar years old. Therefore, radiocarbon years have been calibrated by tree-ring studies, and all published radiocarbon dates are now calibrated and given as calendar years before present (cal. yrs BP) (e.g. Reimer *et al.*, 2004). Care in comparing radiocarbon dates in the older literature with more recent publications is therefore advised. Also, all radiocarbon dates possess error margins, and are always supplied with a ± figure. At the 1σ level (σ is sigma, and 1σ = 1 standard deviation) there is a 68% probability that the age of the sample falls within the date provided, and at the 2σ level (two standard deviations) there is a 95% probability that the date is correct.

The selection of a sample for radiocarbon dating requires in-depth consideration. First, there is the issue of contamination to consider; for example, have the roots from younger plants grown down into the deposit under investigation, or has hard groundwater precipitated older carbon in the form of calcium carbonate in the pores of the sediment to be dated? Second, the derivation of the carbonaceous material should be considered. A peat may be considered **autochthonous**, that is occurring *in situ* and providing an accurate date for the deposit, whereas organic fragments in a sand deposit may be **allochthonous**, that is material transported by currents before being deposited, and therefore the carbon may be much older than the deposit in which it finally becomes incorporated. Considering the reasonably high cost of radiocarbon dating, all these issues need to be addressed.

Because of the relatively rapid decay of ^{14}C, radiocarbon dating cannot be performed on material older than about 60,000 years, and indeed material greater than 20,000 years old is susceptible to yielding erroneous dates. Therefore, in older material alternative dating methods are required. **Uranium-series dating** is similar to radiocarbon dating in that it utilises the radioactive decay of uranium, thorium and radon isotopes to a stable lead daughter isotope. The decay process is considerably longer, so that older uranium-bearing deposits can be dated, including carbonate material such as coral and sea-cave deposits (speleothems) (Edwards *et al.*, 2003).

A now common technique for relatively dating older Quaternary coastal sediments is that of **amino acid racemization** (AAR) (Kaplan, 2003; Clarke and Murray-Wallace, 2006). This requires fossil mollusc shell material of the sort commonly encountered in fossil beach deposits, such as mussels, cockles and winkles, from which amino acids can be obtained. Amino acids persist in shell long after death, but after death convert through time from L-isoleucine to D-alloisoleucine, i.e. the process of racemization. Therefore, the L/D ratio indicates time since death. The now seminal work of Davies (1983), on fossil beaches of the Gower Peninsula (Wales, UK), showed that the AAR technique could distinguish between beach remains deposited during different Quaternary interglacial stages. It must be remembered, however, that this is a relative dating tool only and relies on other methods to assign absolute ages to the L/D ratios.

Dating samples in the absence of organic remains can also now be achieved through a number of techniques, with perhaps the most widely used in coastal studies being various **luminescence dating** techniques (Lian and Roberts, 2006), such as optically stimulated luminescence (OSL) and thermoluminescence (TL) dating. Crudely, these work because crystal sediment particles, such as quartz sand, release luminescent energy when exposed to sunlight (i.e. when lying on a beach or in dunes), but when these crystal particles become buried they begin to store this energy. Therefore, by measuring the amount of energy accumulated in a buried sediment it is possible to estimate when it was last exposed to sunlight. There are some potential sources of error with these techniques, but generally they work well, and dates derived from these techniques correspond well with dates derived from other techniques, such as radiocarbon dating.

5.2.2   The altitude of sea-level index points

Establishing the altitude of a sea-level index point requires accurate surveying using a theodolite, and altitude should be given relative to the national survey datum used in the country in question. For example, in the United Kingdom, the survey datum is known as Ordnance Datum (OD) (Newlyn) (see section 3.2.1 for explanation), and all UK altitudes are given relative to OD. Surveying erosional landforms is fairly straightforward, as long as the surveyed feature is well defined, such as a wave-cut notch or shore platform. The altitudinal range of the feature should be established, such as the top and bottom of a wave-cut notch. This requirement results in vertical features, such as cliffs, sea-caves and arches, appearing to have large altitudinal ranges and so reducing the usefulness of these data in accurately reconstructing sea-level changes.

sea-level index points based on depositional evidence are not only more readily dated than erosional landforms, as discussed in section 5.2.1, but levels within deposits to be surveyed are often very restricted as regards their altitudinal range. For example, the sediment boundaries (contacts) between marine and terrestrial deposits commonly make excellent sea-level index points, and these contacts are often sharp and well-defined so that they can be surveyed down to the millimetre scale if required. In coastal lowlands, sediment coring techniques are often employed to survey the buried sediments and in such cases the altitude of the ground surface is established, from which the depth to the sediment contact is subtracted to give the altitude of the sea-level index point.

Some care is required in many cases in assigning altitudes to depositional sea-level index points. This is particularly true where **isolation basins** are concerned (e.g. Selby and Smith, 2007). Isolation basins are bedrock basins that at one time were submerged or at least inundated by tides, but have since been isolated from the sea by a relative fall in sea level. The altitude of the boundary between the marine deposits, which occur in the bottom of isolation basins, and overlying fresh water/terrestrial deposits, does not necessarily represent the altitude of a former sea level. This is because it is the altitude of the rock lip or sill of the bedrock basin that determines the sea-level altitude that is capable of inundating the basin. Thus, an isolation basin may only be completely isolated from the

Plate 5.2 Sediment coring in an isolation basin at Rumach, west coast of Scotland (UK).

sea when the level of highest astronomical tide falls below the altitude of the rock sill; however, the upper surface of the marine sediments in the basin may be several metres lower than the rock sill altitude (Plate 5.2).

Further altitudinal complications that may arise with depositional sea-level index points include the effect of sediment compaction. When soft sediment is buried by subsequent deposition, the overburden weight of this additional sediment causes compaction, and therefore a lowering of altitude from that at which it was originally deposited (Allen, 1999, 2000b). Compaction, if undetected, can introduce significant altitude errors into a sea-level curve, and therefore measures must be taken to estimate the amount of compaction that may have occurred (see Scientific Box 5.1). Not all sediment types compact to the same degree; indeed Allen (1999, 2000b) suggests that clay, silt and sand deposits suffer minimal post-depositional compaction, but that spongy organic deposits such as peat may be greatly compacted.

5.2.3   Tendency of sea level

This simply refers to whether sea level was rising or falling at a sea-level index point and is most easily applied to deposits. This is established with reference to sediment directly underlying the sea-level index point, so that if sea level has appeared to rise from the underlying sediment to the sea-level index point (e.g. a transition from fresh water peat to marine clay) then a positive tendency is assigned. A negative tendency applies to where sea level appears to have fallen from the underlying sediment to the sea-level index point (e.g. a transition from marine clay to fresh water peat). Sea-level curves constructed without indicating tendency simply show the general trend of sea-level change, whereas when tendency is indicated (by convention with arrows on each point) oscillations in the general sea-level trend can be identified, which may be important in interpreting coastal evolution.

Scientific Box 5.1

Sediment compaction and sea-level curves

For sea-level index points that involve peat, it is essential that any post-depositional compaction is established so that the correct altitude can be assigned. In the Somerset Levels, an extensive coastal lowland in southwest England (Plate 5.3), Haslett *et al.* (1998b) describe a Holocene fresh water peat deposit overlain by marine clay (Fig. 5.6). The peat blankets undulating bedrock topography with an altitudinal range of the upper peat surface spanning 2.22 m. Radiocarbon dating demonstrates the peat surface to be the same age along the length of the investigated transect, indicating that either over 2 m of 'instantaneous' sea-level rise has occurred or that the peat surface was originally horizontal and subsequently deformed by compaction. Palaeoenvironmental analysis of the peat to clay contact indicates that the transition from fresh water peat to intertidal salt marsh was gradual and not 'instantaneous', further suggesting that peat compaction has taken place. In this case, 2.22 m of peat compaction occurred with a maximum clay overburden of just 3.16 m. This is worrying as many previous sea-level curves for the Somerset Levels and elsewhere are based on sea-level index points at peat-clay contacts of 20 m or more depth. Therefore, many early sea-level studies need to be reviewed, and more realistic allowances made for sediment compaction, especially where peat deposits are involved.

Plate 5.3 The coast of the Somerset Levels, an extensive low-lying coastal wetland in southwest Britain, underlain by a thick sequence of Pleistocene and Holocene estuarine and marsh deposits (see Hunt and Haslett, 2006).

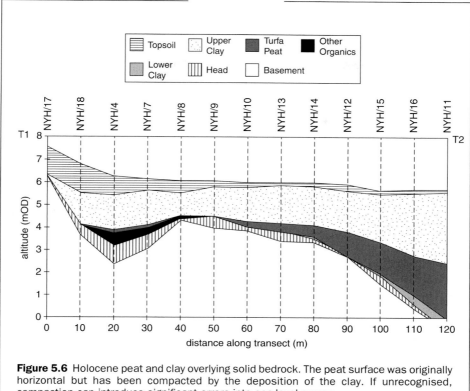

Figure 5.6 Holocene peat and clay overlying solid bedrock. The peat surface was originally horizontal but has been compacted by the deposition of the clay. If unrecognised, compaction can introduce significant errors into sea-level curves.

Source: updated from Haslett *et al.* (1998b, 2001a).

5.2.4   Indicative meaning and range

Indicative meaning refers to the position occupied by a sea-level index point, when it was formed, relative to a reference water level. Usually the reference water level is mean sea level, so that indicative meaning refers to the position of a sea-level index point within the tidal frame at the time. For example, a fossil salt marsh sediment surface may be suitable as a sea-level index point, and may have been deposited at mean high water springs (MHWS); therefore, the indicative meaning of this sea-level index point is MHWS. Indicative range refers to any errors that may be involved in assigning an indicative meaning, so that if it is not certain at what level a sediment was deposited between MHWS and highest astronomical tide (HAT), then the indicative range would be MHWST-HAT.

Establishing the indicative meaning of a potential sea-level index point can really only be achieved through using biological remains (fossils). Many organisms live at, and are restricted to, particular tidal levels. Recovering these tidally specific organisms from either sediments or erosional rock surfaces can often tell us quite precisely the indicative meaning of a deposit or landform. Analysis of diatoms, foraminifera (Technical Box 5.2), amoebae, ostracods, molluscs, crustaceans, pollen and other plant remains have been used with some degree of success in determining indicative meaning (see Haslett, 2002, and Table 5.1).

Table 5.1 The distribution of biological structures in relation to sea level on rocky coasts

Tide level	Zone	Environment	Characteristic structures/ species
high tide	supralittoral zone (also known as the littoral fringe)	rarely submerged, but wetted by spray	lichens and blue-green algae (cyanobacteria)
mean tide	midlittoral zone (also known as eulittoral zone)	regularly submerged by waves and tides	barnacles, algal ridges parallel with the sea surface, hard rim-like accretions of the coralline algae *Lithophyllum lichenoides*, and notch erosion by cyanobacteria and *Patella*
low tide	sublittoral zone (also known as infralittoral zone)	mostly submerged, but upper part exposed at low tide	reef-like constructions built up by vermetid gastropods and coralline algae, but rock is bored by clionid sponges, sea-urchins (*Paracentrotus*) and the bivalve *Lithophaga lithophaga*

Source: Laborel *et al.* (1994) and Pirazzoli (1996)

The main purpose for obtaining information regarding indicative meaning and range is to enable a comparison to be made between different sea-level index points and to construct meaningful sea-level curves. If a group of sea-level index points for a particular region possess different indicative meanings then they must all be standardised before a comparison and sea-level curve can be made. Standardisation is achieved by reducing the points to mean sea level, which is done by subtracting the modern tide level altitude of the coast in question from the altitude of the sea-level index point. For example, take a British macrotidal coast where MHWS is currently 6 mOD and MHWN is 5 mOD, and in coastal lowlands adjacent to the present coast two potential sea-level index points have been established by sediment coring, one with an indicative meaning of MHWS at an altitude of 5 mOD and another with an indicative meaning of MHWN at 4 mOD, both suggesting a positive sea-level tendency with a radiocarbon age of 4000 cal. yrs BP. Although there is a 1 m altitude difference between these two points in the sediment cores, when reduced to mean sea level both points indicate mean sea level to have been −1 mOD at 4000 cal. yrs BP (i.e. 5 − 6 = −1 mOD, and 4 − 5 = −1 mOD). If the indicative meaning for these points were not known, then it may appear misleadingly that an 'instantaneous' 1 m rise in sea level had occurred.

5.3 Coastal responses to sea-level change

Relative sea-level fall and the retreat of the sea from the land is known as a **regression**, whilst relative sea-level rise and the associated inundation of coastal land in known as a **transgression**. Regressions and transgressions lead to the emergence and submergence of coasts, respectively. The manner in which a given coastline responds to such sea-level

Technical Box 5.2

Foraminifera and their use in sea-level and coastal studies

The application of foraminifera to sea-level studies is well established (Berkeley *et al.*, 2007). Salt marsh foraminifera have been shown to occur within specific vertical zones on modern salt marshes that can be related to tide levels and accurately employed as tide-level indicators in palaeo-salt marsh sediments (Gehrels, 2002). Of greatest significance is the recognition that in micro- and mesotidal salt marshes of eastern North America, foraminifera extend up to highest high water, with the highest zone (Zone 1A) characterised by a monospecific assemblage of *Jadammina macrescens*, which extends down to mean highest high water. The recognition of Zone 1A in palaeo-salt marsh sediments is thus a very useful indication of palaeo-tidal level. Haslett *et al.* (1998a), and subsequently Horton *et al.* (1999), have investigated modern salt marshes in the United Kingdom and established foraminifera zonations, which differ slightly from North American counterparts (Fig. 5.7). As well as in temperate salt marshes, foraminifera also appear to be useful in tropical mangrove settings (e.g. Haslett, 2001). Haslett *et al.* (2001b) suggest a model of foraminifera sequences that indicate whether a salt marsh is emerging or submerging under, or in quasi-equilibrium with, sea-level rise through time. The application of this model appears to demonstrate its usefulness (e.g. Pascual and Rodriguez-Lazaro, 2006), although the use of transfer function techniques to reconstruct sea-level change using foraminifera is becoming widespread (e.g. Gehrels, 2000; Massey *et al.*, 2006). Foraminifera are also employed as indicators of coastal pollution (e.g. Nigam *et al.*, 2006) and sediment transport (e.g. Haslett *et al.*, 2000a; Benavente *et al.*, 2005).

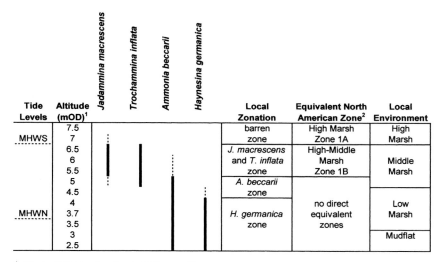

[1] altitude of tidal levels refers to Oldbury-on-Severn
[2] North American zonation based on Scott and Medioli (1986)

Figure 5.7 The relationship of foraminifera distribution to tidal levels and salt marsh environments in the Severn Estuary (UK). The restricted altitudinal ranges make foraminifera excellent tools for reconstructing past sea levels.

Source: updated from Haslett *et al.* (1998a, 2001b).

changes depends on the character of the coast, that is whether it is dominated by depositional or erosional processes. We discussed earlier the relationship between eustasy and isostasy in determining relative sea level (Fig. 5.1); however, along depositional coasts sediment can act as a substitute for, or in addition to, isostasy in certain circumstances. If a beach has an abundant sediment supply, then it may build upwards and advance into the sea regardless of the eustatic regime, a process known as **progradation**. In contrast, if sediment supply is limited and less than that lost from or redistributed within the beach system, through erosion and the action of currents for example, then a beach may retreat landward (i.e. **retrogradation**) under a rising relative sea level, although still maintain its relative tidal position if it is allowed to translate landward (Swift, 1975). Bruun (1962) suggested a model that describes the response of a sandy coast to sea-level rise, where sea-level rise forces the beach to retreat landward through the erosion of the back beach area. This eroded material is then redeposited at the foot of the beach, so that the beach may maintain its equilibrium profile. This model, commonly referred to as the **Bruun rule**, has been widely applied and yet has a number of unresolved problems, such as how can we know when a shoreline is in equilibrium, what timescales are involved in morphological response, and is it fairweather or storm conditions that drive the retreat process (Healy, 1991)? Indeed, some authors suggest the rule should be abandoned (e.g. Cooper and Pilkey, 2004).

The response of different types of coastal systems to sea-level change is often very specific to the given system. Relative sea-level fall usually leads to the emergence, abandonment and fossilisation of a shoreline, and indeed many fossil Pleistocene shorelines are created in this way, such as 'raised' beaches. However, sea-level rise often leads to the landward migration of a mobile shoreline or the submergence and drowning of a stationary shoreline. The response of different coastal systems to sea-level rise during the Holocene and into the future has received much attention, particularly for the following examples of vulnerable and dynamic soft-sediment coastal types.

5.3.1 Salt marshes

The response of salt marshes to sea-level rise depends to a very great extent on the sediment supply into the salt marsh system. The balance between sea-level rise and salt marsh surface elevation, through vertical sediment accretion, is critical (Reed, 1995). If the rate of accretion and surface elevation is less than the rate of sea-level rise then the salt marsh will submerge and drown; if the rate of elevation is higher than that of sea-level rise the salt marsh will emerge and perhaps terrestrialise; the rate of surface elevation may be equal to that of sea-level rise, in which case the salt marsh will neither emerge nor submerge and will exist in equilibrium.

5.3.2 Mangroves

Mangroves were discussed in general in section 3.6.5, but the response to sea-level rise in northern Australia has been examined by Woodroffe (1995). He suggests that mangroves will respond in different ways according to their overall morphology and physical setting. In the tide-dominated Darwin Harbour, mangroves occur fringing the estuary, with many tidal creeks supplying the system with sediment. Therefore, under gradual sea-level rise, this fringing type of mangrove may migrate landward, supplied with sediment from erosion of the submerging lower intertidal zone. Palaeoecological studies suggest that all the studied mangroves migrated onshore in this way during the post-glacial Holocene transgression. However, in other areas that are river-dominated, such as the South Alligator River, extensive deltaic plains of mangroves have been built as the shore has prograded seaward, following the stabilisation of the Holocene transgression, using sediment delivered to the coast by rivers. These near-horizontal plains are susceptible to any future sea-level rise, because they are less likely to be able to migrate landward, and may simply become submerged.

5.3.3 Coastal sand dunes

Foredunes become scarped (Carter and Stone, 1989) as they are reached by the rising sea level, and this releases sediment back into the beach system, perhaps for use in dune building further inland. Sea-level rise may also breach the foredunes to deposit sand behind the foredunes as washover fans. Through both scarping and breaching, sea-level rise promotes the landward migration of coastal dunes. Along many coasts during the Holocene transgression, sand dunes were blown inland ahead of the migrating shoreline. Dune sand blown onto the top of low cliffs and now perched there, offer an opportunity to date this initial sand inundation along a coastline (e.g. Haslett *et al.*, 2000b).

5.3.4 Gravel beaches

Current research on the effect of sea-level rise has suggested that gravel shorelines may respond as follows:

- The crest may build up in height by a transfer of material to the crest from the mid-tide zone as sea level rises. A barrier may start to migrate landward through the process of **rollover**, which is the general transfer of sediment to the landward side of a barrier (Plate 5.4).
- In the absence of adequate sediment supply, the beach will steepen leading to increased reflectivity.
- Crest build-up cannot continue indefinitely, as crest construction requires spilling waves for sediment transport, which in turn requires a measure of dissipative character, so as reflectivity increases the rate of crest build-up will slow, and this shift in morphodynamic domain from dissipative to reflective will induce cusp formation.
- Cusps reaching the crest may then act as a template for overwash and breaching, and the eventual break-down of a barrier.
- In this way barriers will tend to migrate landward under a sea-level rise regime. Beaches confined by backing cliffs may develop swash ramps extending several metres above the high water mark, and with cusp development, ramps may become fragmented and separated by intra-beach erosional bays.

A characteristic feature of a landward-migrating barrier is the presence of a terrace or terraces in the seaward low- to mid-tide zone. These terraces are low-angled and dissipative. Erosional slopes are often formed as a barrier migrates landward. Where a rock platform is exposed, it may supply the barrier with clasts thus slowing down migration, so the areal ratio of barrier to terrace can increase as well as decrease. There is also a type of depositional terrace resulting from sea-level rise and landward barrier migration, which is known as a boulder frame. This frame results from particle sorting, providing a residue or lag of larger and relatively immobile clasts at the foot of the beach. If this residue of clasts remains immobile, then as the barrier migrates landward, the frame will expand in

Plate 5.4 Barrier rollover at Porlock (Somerset, UK): pebbles can clearly be seen encroaching onto agricultural land behind the barrier.

width. With the expansion of low-angled terraces within the barrier morphodynamic system, reflectivity will change from being the dominant domain, to co-dominant to non-dominant through time, giving rise to an overall dissipative domain, which may lead to barrier abandonment. Furthermore, where a barrier progressively loses material to its frame as it migrates, the barrier's migration rate may increase, which may ultimately overstretch the barrier, leading to breaching. In order to maintain stability, therefore, the rate of volumetric loss needs to be balanced by the increasingly dissipative role of the frame (Carter and Orford, 1993).

5.4 Global warming and the threat of future sea-level rise

Global temperatures rose by approximately $0.6°C$ during the twentieth century and climate models estimate that this figure is set to increase up to $6.4°C$ by 2100 (IPCC, 2007a). This global warming has been attributed in part to human activity, and in particular to the burning of fossil fuels that release carbon dioxide (CO_2) into the atmosphere. CO_2, methane (CH_4), chlorofluorocarbons (CFC's), tropospheric (low-level) ozone (O_3), water vapour (H_2O) and some other compounds, are important gases that are able, in the atmosphere, to absorb heat radiated by the earth, whilst allowing the sun's energy to pass through unhindered. Therefore, these gases allow the atmosphere to act like a greenhouse, and are responsible for producing the earth's average temperature of $15°C$. This has given rise to the phenomenon being referred to as the Greenhouse Effect, and without this natural phenomenon, the earth's average temperature would be in the region of $-17°C$. Concern is focused on the increasing levels of CO_2 in the atmosphere from human activity, which is leading to an enhancement of the Greenhouse Effect, resulting in global warming. It must be pointed out that global warming may not be solely due to anthropogenic effects,

---

Scientific Box 5.3

Sea-level rise, coastal retreat and ecosystem impacts

The impact of sea-level rise on habitats is reviewed by Orford and Pethick (2006) with reference to United Kingdom coasts; however, one major ecosystem impact that will affect developed countries is the phenomenon known as '**coastal squeeze**'. This involves coastal systems which respond to sea-level rise by migrating onshore (Pethick, 2001), but are unable to do so because of coastal protection constructions or natural barriers. Pethick (1993) documents such a case from eastern England, an area that is isostatically subsiding, where he recognises extensive estuarine mudflats and salt marshes that are migrating up-estuary as sea level rises. The estuary is lined by sea-walls that act as barriers to the migration of these intertidal environments, and therefore the intertidal zone is becoming narrower as it is 'squeezed' between the rising sea and the sea-walls. This adversely affects intertidal wildlife, such as wading birds, because the intertidal areas, and therefore their feeding grounds, are diminishing. Many important wildlife sites may be lost by such coastal squeeze, but in eastern England some of the sea-walls are being deliberately breached to allow the intertidal area to expand in a landward direction.

---

and that natural phenomena may be contributing, such as variations in solar radiation output.

Regardless of the cause, global warming is unequivocally happening and will continue to affect global sea level. Evidence from tide-gauge stations shows that eustatic sea level has risen by between 12 and 22 cm in the twentieth century (IPCC, 2007a). Four major climate-related factors have been recognised that could possibly explain this rise in global mean sea level. These factors are discussed in the following sections and in Table 5.2, where measurements derived from both tide gauges and satellite altimetry, over different periods, are compared. The rate of sea-level rise resulting from the contribution of these four factors is below the observed rate and is recognised as probably representing an underestimate.

5.4.1   Thermal expansion of the oceans

The volume of the oceans is strongly influenced by the density of sea water, which is inversely related to temperature. Therefore, as the oceans warm up due to global warming, the density of sea water decreases, leading to an increase in water volume, expansion of the oceans, and ultimately a rise in sea level. Sea-level change resulting from density variations is called **steric sea-level change**. Such steric changes may also be brought about by salinity variation which, although it may be locally important, especially at the coast, has a relatively minor effect at the global scale. Estimating oceanic expansion can either be achieved through field observations and recording, or predicted using models. Early research, such as that of Thomson and Tabata (1987) for example, examined a steric height record over 27 years from the northeast Pacific and found that steric heights were increasing at approximately 0.9 mm/year. However, the accuracy of this, and other similar studies, was questioned because interannual variability appeared to produce considerable 'noise' in the record, and also the regional nature of such studies made inferences on a

Table 5.2 Estimated rates of sea-level rise (mm/yr) obtained from tide gauges (1961–2003) and satellite altimetry measurements (1993–2003)

	1961–2003 (tide gauge measurment)	1993–2003 (satellite measurement)
thermal expansion	0.42 ± 0.12	1.60 ± 0.50
glaciers and ice caps	0.50 ± 0.18	0.77 ± 0.22
greenland ice sheet	0.05 ± 0.12	0.21 ± 0.07
antartic ice sheet	0.14 ± 0.41	0.21 ± 0.35
Total	1.1 ± 0.5	2.8 ± 0.7
observed	1.8 ± 0.5	3.1 ± 0.7

Source: Bindoff *et al.* (2007)

global scale problematic. In order to fill information gaps and to overcome problems at the regional scale, the World Ocean Circulation Experiment (WOCE) was designed to provide more accurate and widespread data. With the absence of good observational data, models were used to predict steric changes. Wigley and Raper (1993) used such a model and suggested that for the period 1880–1990, the resultant range of sea-level rise due to thermal expansion of the oceans is in the order of 3.1–5.7 cm. Such research contributed to the figures cited by Warrick *et al.* (1996) for sea-level rise for this period of 2–7 cm. Since that time a number of studies have reported new results for steric height trends (e.g. Antonov *et al.*, 2005). These studies are reviewed by Bindoff *et al.* (2007) who propose estimates of 1.26–2.27 cm for sea-level rise between 1961 and 2003 or 1.1–2.1 cm between 1993 and 2003 based on tide-gauge and satellite altimetry, respectively (Table 5.2).

5.4.2 Melting of glaciers and small ice caps

A number of studies have shown that the majority of non-polar valley glaciers have been retreating since the second half of the nineteenth century, some by several kilometres. This is most likely caused by the rising of altitudinally controlled climate zones found in mountainous regions, so that at the foot of mountain ranges, where many glaciers emerge, temperatures and melting rates are increasing. In the 1980s, Meier (1984) estimated that during the period 1900–61 glacier retreat contributed 2.8 cm, or 0.46 ± 0.26 mm/year, to global sea-level rise. This estimate was then superceded by Warrick *et al.* (1996) who suggested that between 2 and 5 cm has been contributed to sea-level rise in the period 1890–1990. More recently, Bindoff *et al.* (2007) consider that glaciers and ice caps have contributed 1.344–2.856 cm to sea-level rise between 1961 and 2003 and 0.55–0.99 cm between 1993 and 2003 based on tide-gauge and satellite altimetry, respectively (Table 5.2).

5.4.3 Greenland ice sheet

Estimating the contribution of the Greenland ice sheet to sea-level rise over the past century is hampered by a lack of information regarding the dynamics of the ice sheet. It has been suggested that as global temperature increases, precipitation would increase over Greenland, leading to an increase in ice accumulation. Indeed, it was considered that for every 1°C rise in global temperature, precipitation would increase by 4% (Warrick and Oerlemans, 1990). However, such accumulation may affect the interior of Greenland only, as most of the outlet glaciers around the Greenland coast that retreated strongly through the twentieth century continue to do so. Therefore, it is not certain how the Greenland ice sheet has contributed to sea level, and Warrick *et al.* (1996) allow for both negative and positive contributions with an estimate of between –4 and 4 cm, based on a sea-level change rate of ±0.4 mm/year through 1890–1990. The situation appears to be becoming clearer, for although Bindoff *et al.* (2007) estimate a sea-level contribution from the Greenland ice sheet of between –0.294 and 0.714 cm between 1961 and 2003 based on tide-gauge data, figures for the period 1993–2003 from satellite altimetry suggest an entirely positive contribution of 0.14–0.28 cm (Table 5.2). IPCC (2007a) models indicate that after 2100 the Greenland ice sheet may over millennia melt completely under projected levels of climate change and that sea levels may rise by up to 7 m.

5.4.4   Antarctic ice sheet

In a similar manner to the Greenland ice sheet, the Antarctic ice sheet might accumulate ice through an increase in precipitation brought about by global warming. However, unlike Greenland, a number of studies have indicated that the Antarctic ice sheet has a positive balance, that is it accumulates more ice than it loses through melting. For example, Budd and Smith (1985) estimated that of the ice accumulated each year in Antarctica approximately 10% is retained and not lost to melting, resulting in the growth of the ice sheet at a rate in the order of 209×10^{12} kg/year. This positive balance corresponds to a rate of eustatic change of approximately –0.6 mm/year, therefore contributing negatively to global sea-level change. Other studies, however (e.g. Huybrechts, 1990), suggest the contribution to sea level is positive by approximately the same amount. Whatever the contribution, it is finely balanced because the West Antarctic Ice Sheet (WAIS) is grounded below sea level, so that it may be very sensitive to sea-level rise. Sea-level rise may lift the WAIS off the sea-floor by flotation. This would decrease the bottom stresses of the WAIS allowing it to flow faster and thin more rapidly. Ice-berg calving would increase, which would result in a positive contribution to sea-level rise. Warrick *et al.* (1996), with great uncertainty, estimate that the Antarctic ice sheet has contributed –14 to 14 cm to sea-level change during the period 1890–1990. Bindoff *et al.* (2007) present similarly equivocal estimates of the ice sheet's contribution to sea level with –1.134 to 2.31 cm between 1961 and 2003 based on tide-gauge data, and –0.14 and 0.56 cm measured by satellite between 1993 and 2003 (Table 5.2).

5.4.5   Variations in surface and groundwater storage

Surface storage includes water stored in the soil, rivers, lakes and inland seas, an example being the construction of many dams and reservoirs since the 1930s, which have made a negative contribution to global sea-level rise, perhaps by as much as –5.2 cm (Newman and Fairbridge, 1986). Deforestation, however, often leads to water loss from soils through increased surface run-off and evaporation, which constitutes a positive contribution to sea-level rise. Groundwater is water found in porous and permeable rock aquifers, and has been extracted through pumping for domestic, industrial and agricultural purposes. Groundwater pumping amounted to the extraction of over 2000 km³ of water during the twentieth century, which is equivalent to a 0.55 cm rise in sea level (Warrick and Oerlemans, 1990). Warrick *et al.* (1996), again with uncertainty, estimated a contribution to sea level of between −5 and 7 cm during 1890–1990, and Gornitz *et al.* (1997) supported a negative contribution to sea-level rise from these sources. The uncertainty surrounding the contribution from these stores to sea-level rise is still unresolved, with a recent review concluding that the contribution cannot be estimated with much confidence, but is probably less than 0.5 mm/year (Bindoff *et al.*, 2007).

5.4.6   Future sea-level rise and its impacts

The results of a number of model experiments reported by Warrick *et al.* (1996) indicated that a sea-level rise of 49 cm by 2100 is possible, but with a range of uncertainty of 20–86 cm, and represents a rate of sea-level rise that is significantly faster than that experienced over the majority of the twentieth century. Church *et al.* (2001) produced similar sea-level rise estimates of 9–88 cm over the period 1990–2100 with a best estimate of 48 cm. More recent projections of sea-level rise over a similar period, however, present lower estimates in the range of 18–59 cm, but due to the nature of the uncertainties involved in their computation, are unable to provide a best estimate (Meehl *et al.*, 2007). The various contributors to sea-level change in the period 1980–99 to 2090–99 have been assessed individually (Table 5.3).

Projections given by Warrick *et al.* (1996) and Church *et al.* (2001) showed that throughout the twenty-first century, even under scenarios of lower greenhouse gas emissions, sea level continues to rise. This suggests that even if greenhouse gas emissions were cut completely now, sea level would continue to rise. This is mainly because of a lag effect caused by the ability of the oceans to store heat, in what is called thermal inertia. This means that society is committed to a sea-level rise. However, global sea-level rise will have different impacts on different coastlines, mainly due to isostatic movements, but there is considerable concern for the majority of the world's coasts (see Management Box 5.4). Concern is not centred on the day-to-day increase in tidal range, which can in most cases be contained, but on the increased ability of extreme events, such as storm surges and very high tides, to flood coastal lowlands and cause extensive coastal erosion. Deep storm depressions are predicted to become more frequent under global warming, so that increased sea levels, coupled with surge conditions and high precipitation in coastal lowland river catchments, are inevitably going to mean higher frequency and higher magnitude flood events.

Table 5.3 Projected global sea-level rise (SLR) between 1980–99 and 2090–99 and estimates of the rate of SLR during 2090–99. The ranges comprise output from six different climate change scenarios

	Projected SLR (m) 1980–99 to 2090–99	Rate of SLR (mm/yr) 2090–99
thermal expansion	0.1 to 0.41	1.1 to 6.8
glaciers and ice caps	0.07 to 0.17	0.5 to 2.0
greenland ice sheet	0.01 to 0.12	0.2 to 3.9
antartic ice sheet	−0.14 to −0.02	−2.7 to −0.3
Total	0.18 to 0.59	1.5 to 9.7

Source: Meehl *et al.* (2007)

5.4.7 Managing global sea-level rise

Carter (1988) realised that predicted sea-level rise into the twenty-first century will lead to coastal erosion, redistribution of sediments, wetland submergence, floodplain water-logging, and salt contamination of coastal aquifers. Disruption will be caused to residential, industrial and commercial activities, transport routes will be severed, and agricultural land ruined. Thus, with global sea-level rise there is going to be an increasing demand for management solutions to protect thousands of kilometres of coast, and addressing these consequences will require a major redistribution of money at both the national and international levels. Good coastal protection schemes, such as the Thames Barrier in London, are not viable on a worldwide basis, and governments need to accept the loss of

---

Management Box 5.4

Socio-economic impacts of future sea-level rise

Reports of the Intergovernmental Panel on Climate Change (IPCC) outline the socio-economic implications of predicted sea-level rise due to an enhanced Greenhouse Effect (e.g. Bijlsma, 1996; Nicholls *et al.*, 2007). These include the following.

- There will be general negative impacts on various coastal amenities and activities, such as tourism, fresh water supply and quality, fisheries and aquaculture, coastal agriculture, human settlements, financial services, and human health.
- Coastal wetlands will come under pressure as they become squeezed by rising sea level against landward barriers and/or become starved of sediment.
- Protecting low-lying islands, such as the Maldives and Marshall Islands, and countries with extensive low-lying deltaic plains, such as Bangladesh, Nigeria, Egypt and China, will probably be extremely costly.
- The numbers of people globally who currently suffer from coastal flooding (40 million) could double or triple by 2100, especially from the 2080s onward in the mega-deltas of Africa and Asia.
- Attempting to adapt to sea-level rise may mean that some societies will face difficult decisions regarding which coasts to protect and which to abandon, and tradeoffs between environmental, economic, social and cultural values will have to be made.
- Coral reefs may also suffer increased bleaching episodes as sea-surface temperatures rise by 1–3°C.

Furthermore, in an early estimate, Houghton (1994) suggested that for the United States alone, the financial cost of coastal protection and land loss will amount to seven thousand million US dollars, which doesn't include the cost from sea water contamination of water supplies, agricultural losses, and health problems that may be caused by sea-level rise. This figure is approximately 1% of the US gross domestic product (GDP), and a similar percentage is predicted for other developed countries, whilst for some developing countries costs could be as much as 6% of GDP. IPCC (2007b) supports this early view, indicating global mean losses from a 4°C rise in temperature would be in the range of 1–5% of GDP.

coastal land. The early anticipation of sea-level rise is of great value because it allows planning to be introduced gradually.

Various national organisations have instigated policies and plans to address the sea-level rise management issue. Nicholls *et al.* (2007) review the principal adaptation practice options available, which are being considered in a number of coastal management strategies. For example, Bray *et al.* (1997) provide advice to coastal management decision-makers and suggest a number of planning options for the coast of southern England:

- *Retreat* – this has a number of meanings, including the practice of **managed retreat** or realignment. It may involve the relocation of coastal communities or industry, with a prohibition on further development. This would require a strong governmental role with supportive legislation, and with the associated expense of relocation costs and compensation. Coastal processes would operate naturally in response to sea-level rise, but realistically only undeveloped, low-density or precarious land would be left to retreat in this way. Alternatively, retreat may involve 'conditional development' whereby the coast continues to be used, but with conditions in leases set by government that will eventually enable the sea to reclaim the area. The final retreat option involves the surrendering of governmental responsibility: after stating the risks, government abandons the coast to private market forces.
- *Accommodation* – this essentially means learning to live with sea-level rise and coastal inundation. This may be through an adaptation to the effects of sea-level rise, such as elevating buildings, enhancing storm and flood warning systems, and modifying drainage. For communities already used to coastal flooding, or where retreat or protection are not socio-economically possible, this may be an ideal option, but it does require high levels of organisation and planning, and community participation, all of which can be quite costly to implement. It may be possible to change activities, such as changing agricultural practices, to suit the new environment, or simply to accept the risks of inundation, and accept increasing insurance premiums. This final option is only really viable in urban areas perceived to be at low risk from coastal inundation.
- *Protection* – this involves the physical protection of the coast, such as the construction of 'hard' structures (e.g. sea-walls, groynes), for developed stretches of the coast that are particularly vulnerable to erosion or inundation. A disadvantage of 'hard' protection is that it is fixed, and if the shoreline does change, a new line of defences would have to be built, representing a duplication of the original considerable cost. The alternative method of protection is known as the 'soft' option, which involves artificial beach nourishment, for example, which is commonly employed, ironically, to afford a measure of protection for 'hard' structures. These 'soft' options often mimic natural processes, but their maintenance can be costly, especially if sediment transport removes introduced sand from the beach system.

The option adopted by managers and planners of any given coast will be influenced by human factors, such as population, settlement, business and industry (i.e. socio-economic), more than physical factors. Figure 5.8 illustrates schematically the relationship between the value of a coast to society (which may not always refer to its economic value) and the cost of protecting that coastline. It indicates that where the value of the coastal land is less than the cost of protecting it, then communities will either have to accommodate the effects of sea-level rise, or accept retreat.

It has also been suggested that it may be possible to offset global sea-level rise by transferring sea water onto land. Candidates for this flooding include the Dead Sea region in the Middle East, which lies well below present sea level, with the advantage that hydro-electric power could be generated in the water transfer process from sea to land (Carter,

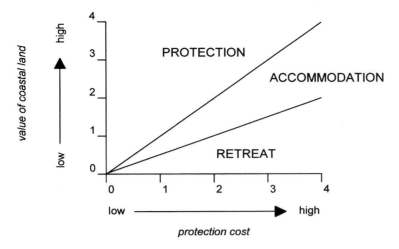

Figure 5.8 The relationship between the value of coastal land and the cost of protecting that coastline (the numbers on the axes are intended to indicate comparable value and cost, although the value of coastal land may not always refer to its economic value).

1988). However, schemes like this are perhaps not likely to reach fruition due to local socio-economic and political conditions.

Summary points

- Sea level is not constant in either space or time.
- Relative sea level represents the balance between eustasy and isostasy, both of which are strongly influenced by the growth and decay of continental ice sheets through the Quaternary.
- Sea-level curves depicting sea-level change are best constructed using sea-level index points for which age, altitude, tendency of sea level, and indicative meaning and range are known.
- The repsonse of coastlines to sea-level change varies, but landward migration is likely for depositional coasts where the sediment is mobile and/or in abundant supply. Immobile coasts become submerged and drown.
- Sea level appears to have risen throughout the twentieth and early twenty-first centuries and is predicted to continue to 2100 and beyond as a consequence of human activity and the enhancement of the Greenhouse Effect.
- As a consequence of predicted sea-level rise, coastal societies must either (a) learn to accept the loss of coastal land, (b) accommodate sea-level rise, or (c) protect coastlines.

Discussion questions

1. Discuss the method of construction, and evaluate the accuracy of Holocene sea-level curves.
2. A number of climate-related factors are recognised by the Intergovernmental Panel on Climate Change that could possibly explain a rise in global mean sea level on the 100-year scale. Examine the sensitivity of these factors to changes in climate, and outline how they have stimulated sea-level change over the last 100 years.
3. Examine the variety of options open to coastal populations in response to future sea-level rise.

Further reading

See also

The tectonic classification of coasts, section 1.3.2
Coastal systems and sea-level examples, sections 2.4, 2.5.1, 3.5.2, and 4.7
Coastal management issues, section 6.2

Introductory reading

Ice Age Earth: Late Quaternary Geology and Climate. A. G. Dawson. 1992. Routledge, London, 293pp.
An extensive review of climate and environmental change from the last glacial onwards.

Sea levels: change and variability during warm intervals. R. Edwards. 2006. *Progress in Physical Geography*, **30**, 785–796.
A useful update on sea-level research.

Climate and Sea-level Change: Observations, Projections, and Implications. R. A. Warrick, E. M. Barron and T. M. L. Wigley (eds). 1993. Cambridge University Press, Cambridge, 424pp.
A collection of papers examining the issue of global climate change and associated sea-level rise.

Present-day sea-level change: a review. R. S. Nerem, E. Leuliette and A. Cazenave. 2006. *Comptes Rendus Geoscience*, **338**, 1077–1083.
A brief review of observed sea-level change for the last 50 years.

Impacts of Sea-level Rise on European Coastal Lowlands. S. Jelgersma and M. J. Tooley (eds). 1992. Blackwell, Oxford, 267pp.
An example of how sea-level rise impacts on a regional scale.

Advanced reading

Sea-level Research: A Manual for the Collection and Evaluation of Data. O. van de Plassche (ed.). 1986. Geo Books, Norwich, 618pp.
The standard reference work for the collection and interpretation of sea-level data.

Quaternary Environmental Micropalaeontology. S. K. Haslett (ed.). 2002. Arnold, London, 340pp.
A thorough overview of the use of microfossils in the study of Quaternary palaeoenvironments, including sea-level research applications.

Sea-Level Changes: The Last 20 000 Years. P. A. Pirazzoli. 1996. Wiley, Chichester, 211pp.
A well-written introduction to sea-level changes on a geological timescale.

Sea Level Rise: History and Consequences. B. C. Douglas, M. S. Kearney and S. P. Leatherman. 2000. Academic Press, New York, 232pp.
Examines historic sea-level rise and is accompanied by a CD-Rom containing historic sea-level data.

Submerging Coasts: The Effects of a Rising Sea Level on Coastal Environments. E. C. F. Bird. 1993. Wiley, Chichester, 184pp. A useful review of the response of coastlines to rising sea levels.

6 Coastal management issues

Coasts are some of the most populated areas in the world, and the interaction between humans and the environment often throws the natural coastal system out of equilibrium. Management is required to mitigate against these human impacts on the environment and upon society. This chapter covers:

- **sustainable coastal-zone management**
- **a range of coastal management issues, such as population growth, coastal uses, coastal hazards and administrative issues**
- **a number of approaches to integrated coastal-zone management**
- **various examples of human impacts upon the coastal environment**

6.1 Introduction

Hopefully this book, if nothing else, has imparted an appreciation of the sensitivity of coastal systems, and that intervening in one part of a system will have an effect on the whole system. Human activity at the coast invariably intervenes in coastal systems, often with detrimental impacts. In order to minimise these impacts, which affect both the natural environment and the intervening human population, coasts must be managed. Management is also concerned, however, with non-anthropogenic processes that affect coasts, but impede on human activity, such as wave attack and coastal erosion. Therefore, the aims of coastal management are to facilitate the human use of the coastal zone, but minimise the impacts of such human use, and to protect human interests at the coast from natural and human-related processes.

Sustainability is now a key word in environmental management, so that all human use of the coast should meet the needs of the present population without jeopardising the opportunities of future generations, so that our descendents will be able to use the coast in much the same way as we do today. Sustainable coastal management therefore means that all human activities should be non-destructive, and that resources we exploit should be renewable. Against such a benchmark, it is clear that many practices are not sustainable and that our descendents will not inherit the opportunities that the present generation enjoys. However, we are at a critical time for the management of the global environment, and if sustainable management practices are widely adopted now, the opportunities our children inherit will be maximised.

6.2 Coastal management issues

The issues that coastal management is expected to address are very diverse and mostly local or regional in extent. However, some coastal management issues do affect the global coastline, such as the threat of future sea-level rise as a result of global warming, as discussed in Chapter 5. Nevertheless, all these issues must be appreciated by the coastal manager so that holistic responses can be made. This, of course, is in addition to an intimate understanding of the operation of coastal systems, which has been the main focus of this book. Kay and Alder (2005) discuss at length the variety of coastal management issues and employ many excellent case studies. The following sections examine these issues.

6.2.1 The growth of coastal populations

The global population passed the six billion mark before the end of the twentieth century, and over 50% of these people live in the coastal zone – that is equal to the entire global population of the 1950s. The future predictions are just as alarming in that by 2025 more people will live in the coastal zone than were alive in the 1990s. This incredible pressure placed upon the coast by an ever-increasing global population is the principal reason why coasts are under threat, and why they need the level of management and protection that they do.

This increase in coastal populations is not restricted to developing countries, with some 50% of people in developed countries living less than 60 km from the coast. The reasons for this population concentration at the coast are complex, but include the following:

- The general global population is increasing, with indigenous coastal populations growing.
- Rural to urban migration: rural depopulation is currently of global concern, and because many of the world's major cities are sited on the coast, they are attracting people from displaced rural populations.
- Inland to coastal migration: this concerns the migration of people from inland areas specifically and intentionally to the coast for economic, social and recreational reasons. In developed countries this is often brought about by people retiring to the coast.

The human geography of coastal cities is of significance in understanding coastal population growth, in that most major coastal cities have developed as a major port. This allows the import and export transport of material and goods, and encourages industry and business to develop at or near the port site. Such ventures stimulate employment and investment opportunities, both of which attract people to come and work and live in the coastal city. Under these circumstances, expanding coastal populations put immense pressure on local planning authorities to provide suitable land for urban and further industrial development, so consuming more coastal land. This can form the basis of a cycle which leads to the continuous increase in city size. Indeed, some neighbouring coastal cities have been known to merge in this way, such as Tokyo and Osaka in Japan (Kay and Alder, 2005).

Management Box 6.1

The impact of development in the Florida Keys (USA)

The Florida Keys are an extensive coastal wetland in the southern USA. The coast comprises barrier islands, mangroves, sea-grass beds and coral reefs. The area is a popular retirement region, due to its generally good weather, easy-going lifestyle, and diverse recreational activities. Tourism is also a growing industry here. Development of the Keys has expanded greatly in recent years so that some mainland and barrier island shorelines are completely given over to residential development, with up to 2500 people per km^2 (Finkl and Charlier, 2003), and little of the natural shore remains. Deford (1999) outlines some of the impacts that this continued residential development and recreational activity has had on the mangrove, sea-grass and coral reef environments:

- *Mangroves* – these are extremely important to the Florida Keys marine environment, as they act as nurseries for many species of fish and spiny lobster, their roots filter pollutants out of the sea water, and they serve to buffer the coast against hurricanes. Although protected by law, mangroves are still being deforested by landowners and developers in order to improve waterfront views and to create docks for pleasure craft.
- *Sea-grass beds* – these occur in shallow water seaward of the shoreline and are important in trapping sediment which may otherwise be transported further seaward where it would be harmful to coral. They also absorb excess nutrients in the water from poorly treated sewage and storm water runoff from urban areas. However, if nutrient levels are too high, algal blooms occur that block off sunlight reaching the sea-grass, killing them (Barile, 2004). Pleasure boat propellers create grooves in the sediment which are difficult for sea-grass to recolonise, and similar effects are produced by anchors and lobster pots dragged across the beds. Dredging of channels undermines sea-grass roots, so promoting localised erosion and releasing sediment which requires further dredging.
- *Coral reefs* – these are important in dampening the effect of waves at the shoreline, protecting the landward sea-grass beds and mangroves. And yet they suffer physical damage from anchors, grounded vessels, and from souvenir collecting. Pollution affects the reefs through algal blooms limiting light and reducing the activity of the photosynthesising coral symbionts, thus increasing disease and coral mortality.

The Florida Keys are managed by the Florida Coastal Management Program (FCMP) which operates under the Florida Coastal Management Act of 1978. The FCMP administers the Florida Keys National Marine Sanctuary, and contributes towards schemes set up to help mitigate against the threat of hurricanes, and to emplace hurricane evacuation plans (Kay and Alder, 2005).

6.2.2   Use of the coast

The use of the coastal zone is probably as old as our species and comprises extremely diverse practices. Besides the use of coastal land for industrial and residential development (discussed in section 6.2.1), coastal use may be treated within four categories: resource exploitation, infrastructure, tourism and recreation, and conservation (Kay and Alder, 2005).

6.2.2.1 Resource exploitation

This includes the use of the coast's physical and biological resources. Although ideally such exploitation should be done on a sustainable basis, this has not generally been the case. Coastal fish stocks have declined with overfishing, so that many species are depleted or indeed fully fished. The relatively recent rise of aquaculture on a large scale, for fish, prawns, seaweed and sea cucumbers, has resulted in the loss of natural coastal wetland for pond creation, and major pollution problems from the high levels of nutrient output (e.g. unconsumed food, faecal pellets) often leads to coastal eutrophication, especially where the aquaculture is based on caged stock in the coastal sea. Such aquaculture may also lead to the increase of disease in marine organisms, and the inevitable accidental release of exotic species may displace native populations.

Mangrove deforestation is also a major problem for tropical coastlines. Traditionally mangroves have been used for firewood and construction material, but the present high demand placed upon these resources by growing coastal populations is becoming unsustainable. Also, major tracts of mangroves are being converted to aquaculture. Uniquely, 50% of the mangrove forests of the Mekong Delta were destroyed during the Vietnam War by the US Army's use of chemical defoliants and napalm (Viles and Spencer, 1995). The impact of this activity is to reduce coastal biodiversity and sedimentation, and to leave the coastline more vulnerable to cyclone events and flooding.

The exploitation of oil and gas, and of other materials that are mined in the coastal zone is entirely unsustainable, as it involves the permanent removal of resources. Deltas are often rich in hydrocarbon reserves and many major oil and gas fields are situated on or near deltaic systems, such as the Mississippi Delta in the southern United States. The construction of oil rigs and pipelines is often large-scale and involves degradation of the coastal zone. Mining of coastal sediment through quarrying and dredging is practised for supplying aggregate for the building industry (Plate 6.1), for deepening shipping channels and for supplying sand for beach nourishment schemes.

6.2.2.2 Infrastructure uses

This refers to the building and expansion of ports, with associated shipping, industrial and urban development, including road and bridge building (Plate 6.3). With the advent of very large ocean-going vessels, ports have had to respond by deepening shipping channels and providing deep-water docking facilities. The ability to transport increased cargo loads has stimulated industrial expansion and the development of major urban centres at the coast. This has led to the building of roads and bridges to facilitate the distribution of imported and manufactured goods, and to allow exports to reach the port. The demand for coastal land for development in these situations is great, and has prompted coastal wetland reclamation and even the construction of artificial land. Pollution is a major problem in ports, with some of the worst pollution events involving oil spills from foundered tankers. Unfortunately, there are many examples of this, such as the Sea Empress disaster in the port of Milford Haven (UK) in 1996.

Plate 6.1 Quarrying of dune sand and gravel in St Ives Bay, Cornwall (UK) in 1996.

Management Box 6.2

The impacts of sediment extraction

Sediment extraction for aggregates may destabilise coastal systems and lead to long-term shoreline retreat. For example, in the Baie d'Audierne (Brittany, France) the shoreline has retreated over 50 m since about 1945. Plate 6.2 clearly shows the position of a German World War II gun emplacement presently lying in the middle of Plage de Tronoan in the southern part of the bay. Guilcher *et al*. (1992) describe how these military installations were originally located along a high gravel barrier which controlled the recession of the coastline. The gravel barrier suffered from sediment extraction both during the war and afterwards, which left it susceptible to breaching by waves. Sand was washed through breaches during storms to form deltas over 200 m wide on the landward side of the barrier. Onshore winds transported this sand further inland to form dunes which rapidly became stabilised by vegetation. In this case, there has been not only extreme coastline retreat, but also a change in the coastal environment. There is now no sign of the former gravel barrier at Plage de Tronoan and the narrow belt of low dunes present today is showing signs of continued retreat.

Plate 6.2 Plage de Tronoan in the Baie d'Audierne (Brittany, France) in 1994. The German army built the blockhouse on the beach during World War II on top of a gravel barrier. Gravel extraction during World War II and after led to the destruction of the barrier, retreat of the shoreline and construction of a dune system.

Plate 6.3 Infrastructure use of the coast is well illustrated by the view of the port of Penzance in Cornwall (UK). The area is well built-up and very little undeveloped coast remains in the immediate vicinity.

Management Box 6.3

The problem of coastal litter

It is almost impossible to visit a beach now without seeing some litter (Plate 6.4). Many coastlines around the world are plagued by litter, and this is posing great problems to health, scenery, wildlife, and the tourist industry. Many municipal and voluntary schemes try to clear beaches of litter, but it appears to be a losing battle for fresh litter usually returns upon each high tide. It is now widely appreciated that in order to tackle the problem of beach litter, the origin must be identified, so that it may be stopped at the source. There are three potential sources for this litter:

Plate 6.4 Strandline deposits of litter on Plage de Mezpeurleuch (Brittany, France) – an all too familiar sight on beaches today.

1. *Marine sources* include rubbish thrown overboard by boats and ships. This is often the greatest source of foreign litter found on beaches.
2. *Beach sources* refers to direct littering onto a beach, most often by beach-users, mainly tourists.
3. *Riverine sources* comprise litter washed into the sea from rivers, and may be from sewage outlets and fly-tipping (illegal disposal of waste) on river banks.

Williams and Simmons (1997) have analysed litter on beaches in the Bristol Channel (UK) and found some very alarming amounts of litter, the majority of which they attribute to discharge into the sea from rivers, such as the River Taff. For example, Merthyr Mawr beach was found to contain 550 plastic bottles, 75 sanitary items, and 210 cans per kilometre of strandline. Also, in Australia, Cunningham and Wilson (2003) found over 2500 items of litter per kilometre of beach around Sydney, with the vast majority (90%) of it comprising plastics either left by beachgoers or brought inshore during storms. They argue that this global problem requires immediate attention.

6.2.2.3 Tourism and recreation

Tourism is perhaps one industry which could, if managed properly, be compatible with environmental sustainability. However, it is often poorly managed, both in developed and developing countries, and leads to the degradation of coastal sites, reducing their suitability as tourist destinations and resulting in economic decline. Tourist facilities, such as caravan parks (Plate 6.5), camping sites, resorts, golf courses and marinas, are often installed without full appreciation of coastal systems. Common impacts of tourist and recreation activity include the following:

*   Coastal sand dune degradation occurs where caravan and camp sites are built behind dunes, and where pathways cross the dunes which incise with continued use, leading

Plate 6.5 Caravans and motor-homes parked on a salt marsh at Mont St Michel (Normandy, France), the weight of which is affecting marsh cliff stability.

to the undermining of dune vegetation and blowout formation. Horse riding and on-dune driving can also have the same effect. This may ultimately result in the fragmentation of the dune system, leaving the land behind the dune vulnerable to inundation during storms.

- Heavily used coastal cliff-top footpaths, if not properly managed, can erode causing slope instability and soil degradation (Plate 6.6).

Plate 6.6 Pointe du Raz in Brittany is the westernmost tip of mainland France and suffers intense tourist pressure, culminating in the summer with 2000 visitors per hour, which has had serious environmental impacts, especially soil erosion. Measures taken to combat these impacts has taken the form of building the visitor centre over 1 km away from the site, providing designated paths and restricting access to certain areas. (a) Photograph taken in 1997 shortly after the start of the rehabilitation programme; (b) photograph taken in 2004 indicating the level of vegetation regrowth that has occurred

Source: 2004 photograph kindly supplied by Dr Heather Winlow, Bath Spa University.

- Artificial marinas can provide new quiet water conditons that act as a sink for sediment deposition, therefore removing sediment from longshore drift systems, which may leave downdrift sections of coastline vulnerable.
- Water craft can damage the sea-bed; for example outboard motor propellers damage sea-grass beds, and the deployment of anchors damages coral.
- The outflow of sewage into the sea is poorly treated in many developing countries, if at all. This pollution causes environmental damage through eutrophication, and poses severe health risks for the tourist population.

These are just some of the potential impacts of tourist activities on the coast, and do not include the construction of tourist accommodation which can consume large tracts of coastal land.

Scientific Box 6.4

Coastal eutrophication

Eutrophication occurs when excess nutrients enter the aquatic environment and cause algae to bloom, which in turn depletes oxygen in the water. The types of nutrients that may trigger eutrophication include nitrates and phosphorus, and these may be sourced from agricultural fertilisers, poorly treated or untreated seawage, and animal wastes. In the coastal waters of Brittany (France) the problem is quite severe, as it is an intense agricultural region. Blooms of *Ulva* algae occur each year in sheltered embayments, such as the Baie de Douarnenez, and become deposited on beaches (Plate 6.7), the sight and smell of which seriously detracts from the

Plate 6.7 Vast accumulation of algae on Fautec Beach (Brittany, France), an unsightly consequence of coastal eutrophication.

tourist value of the locations. However, it must be remembered that not all algal blooms are the result of eutrophication (Sellner *et al.*, 2003).

Where coastal eutrophication does occur, algal deposits are collected (often daily in tourist seasons) to clean beaches, and re-used as agricultural fertiliser. In 1992, 82,000 m^3 were collected from Breton beaches (Fera, 1993), which had decreased to 51,000 m^3 by 1998 (Morand and Merceron, 2005). Furthermore, coastal sedimentation is important as a sink for pollutants, especially mining wastes, radionuclides and trace metal contaminants. However, these sinks are often only temporary, and pollutants can be disturbed and re-released by dredging activity and boat traffic, as well as by natural processes of erosion. In a number of cases, uncontrolled discharge of pollutants has led to the complete elimination of bottom-living organisms.

6.2.2.4 Conservation uses

There are many nature reserves and wildlife parks situated at the coast. However, Agenda 21 of the Rio Earth Summit of 1992 requires that further measures be taken to safeguard global biodiversity. Kay and Alder (2005) suggest that Agenda 21 requirements from the coastal zone will be met through the planning of new conservation areas. Indeed, new coastal developments must often address the conservation issue, and provide 'buffer' zones which act as transitional spaces from developed to natural spaces, that allow physical processes to operate. The creation of new reserves may often simply involve the protection of natural coastal environments. There are cases, however, where the conservation interest in a coastline is at odds with the operation of the natural system. This is particularly so for waterfowl reserves in coastal wetlands, where grazing, nesting and feeding habitats are needed, so that some areas should be permanent water, and others semi-permanent marshy grassland. In order to obtain this landscape diversity in a coastal wetland, the water table must be controlled, embankments built around some areas for pasture, and drainage ditches dug. Also, nature reserves stimulate nature-based recreational tourism, which itself may be perceived as degrading the coastal environment (e.g. Priskin, 2003).

6.2.3 Coastal hazards

Coastal hazards are important management issues as it is this group of issues that cause obvious environmental and social disasters. These include short-term events, such as the landfall of cyclones, hurricanes and typhoons, and the impact of tsunami, and longer-term issues such as wave activity and the predicted gradual rise in future sea level that may be associated with climate change (see Table 6.1 for a summary of coastal protection measures). Managing the coast to combat these issues often requires high levels of finance and long-range management plans. For example, combating flooding associated with storm surges has involved the construction of costly barrages across estuaries, such as the Thames Barrier in London (UK). In Japan, where tsunamis occur relatively frequently, many coastal towns have built high walls and gates around key areas that will act as a retreat if a tsunami approaches. In a similar manner, elevated platforms are constructed along some vulnerable low-lying coasts so that people can escape above the flood waters. The management responses to predicted sea-level rise have already been discussed in section 5.4.7. In addition to these extreme events, above-average wave activity can accelerate coastal erosion and cause cliff retreat, destroying cliff-top buildings.

Table 6.1 A summary of engineered coastal protection measures

management issue	engineered solution(s)	types	description	problems	illustration
cliff erosion	seawalls	vertical wall	a wall constructed out of rock blocks, or bulkheads of wood or steel, or simply semi-vertical mounds of rubble in front of a cliff	rock walls are highly reflective, bulkheads less so. Loose rubble however, absorbs wave energy	
		curved wall	a concrete constructed concave wall	quite reflective, but the concave structure introduces a dissipative element	
		stepped	a rectilinear stepped hard structure, as gently sloping as possible, often with a curved wave-return wall at the top	the scarps of the steps are reflective, but overall, the structure is quite dissipative	
		revetment	a sloping rectilinear armoured structure constructed with less reflective material, such as interlocking blocks (tetrapods), rock filled gabions, and asphalt	the slope and loose material ensure maximum dissipation of wave energy	
coastal inundation	seawalls	earth banks	a free-standing bank of earth and loose material, often at the landward edge of coastal wetlands	may be susceptible to erosion, and overtopped during extreme high water events	
	tidal barriers		barriers built across estuaries with sluice gates that may be closed when threatened by storm surge	extremely costly, and relies on reliable storm surge warning system (e.g. Thames Barrier)	
beach stabilisation	groynes		shore-normal walls of mainly wood, built across beaches to trap drifting sediment	starve downdrift beaches of sediment	
	beach nourishment		adding sediment to a beach to maintain beach levels and dimensions	sediment is often rapidly removed through erosion and needs regular replenishing; often sourced by dredging coastal waters	
offshore protection	breakwaters		structures situated offshore that intercept waves before they reach the shore. Constructed with concrete and/or rubble	very costly and often suffer damage during storms	
tidal inlet management	jetties		walls built to line the banks of tidal inlets or river outlets in order to stabilise the waterway for navigation	the jetties protrude into the sea and promote sediment deposition on the updrift side, but also sediment starvation and erosion on the downdrift side	

6.2.4   Administrative issues

The diversity of problems facing the coast, and the varying temporal and spatial scales at which they operate, inevitably brings in many different organisations with an interest in the management of a coastline. These may typically include local, regional, national and international administrative authorities, such as councils and governments, environmental and conservation organisations, and other groups which may represent residents, tourist and industrial interests. It is not unusual along any coastline for a mosaic of different coastal managers to exist. There is a strong argument, however, that for effective management of a coast, an integrated approach should be adopted, either by all small concerns working closely together to decide widely applicable management plans, or by amalgamation, so that one body has overall responsibility for management planning.

Cicin-Sain (1993) outlines a continuum of coastal management integration (Kay and Alder, 2005). Five stages along this continuum have been described:

1. *fragmented stage* – many small organisations are operating independently with little communication between them;
2. *communication stage* – there is regular but occasional communication between different organisations;
3. *coordination stage* – different organisations are in close communication and synchronise their work;
4. *harmonisation stage* – different groups synchronise their work along universally agreed policy guidelines;
5. *intergration stage* – there is complete synchronisation and formal mechanisms along which work must be carried out, therefore independence of individual organisations is lost and fully integrated coastal management is achieved.

As one moves towards integration along this continuum there is usually a high degree of legislation, that allocates management responsibility to one particular organisation, which may or may not be governmental.

6.3 Approaches to coastal-zone management

From section 6.2 it is clear that management of the coastal zone operates at many different levels, from local to national. It is often the financial resources, and legal authority and responsibility at these different levels that determine the form of management employed. With such a diversity of management approaches practised, it is impossible here to detail all, therefore only a small number of approaches are reviewed to serve as examples.

6.3.1   SCOPAC – coastal zone management in southern England

In Britain, the Coastal Protection Act of 1949 states that the protection of the coastline from erosion is the responsibility of the local authority, and protection from flooding is within the remit of the Environment Agency. Therefore, coastal management in Britain has been very fragmented. In 1986, the local authorities of the southern England coast realised the need for closer cooperation between neighbouring authorities. This realisation came in the light of increasing shoreline erosion problems and the *ad hoc* nature of local management solutions, which often had adverse effects along an adjacent authority's coast. As a result, the Standing Conference on Problems Associated with the Coastline

(SCOPAC) was created, and included the coast between Dorset in the west and West Sussex in the east. Membership includes officers and elected members of district and county councils, the Environment Agency, harbour authorities and English Nature, the national nature conservation body. The full conference meets three times a year, local authority officers at least another three times, and specialist subgroup meetings also occur (Hooke and Bray, 1995; Carter *et al.*, 2000).

SCOPAC has a number of aims and objectives (Hooke and Bray, 1995: 359):

- 'to ensure a fully coordinated approach to all coastal engineering works and related matters between neighbouring authorities . . .
- to eliminate the risk of coastal engineering work carried out by one authority adversely affecting the coastline of a neighbouring authority.
- to exchange information of the success or failure of specific types of coastal engineering projects.
- to establish close liason with government and other bodies concerned with coastal engineering projects.
- to identify aspects of overall coastal management where further research work is required and to promote such research.'

In order to facilitate the last of these aims, each local authority makes financial contributions to SCOPAC for the funding of research. Some of the funds are also used to influence coastal management policy decisions at the governmental level, and have even included the organisation of field trips to which members of parliament have been invited to observe coastal problems at first hand. SCOPAC has been successful in fulfilling most of its aims, and this has prompted other authorities around the British coastline to create local groups. The setting up of these coastal management groups has been done, as far as possible, according to the distribution of natural coastal cells, that is stretches of coastline which behave as essentially self-contained coastal subsystems, with identifiable inputs, transfers, stores and outputs of sediment (Hooke and Bray, 1995). By delineating group boundaries coincident with natural systems' boundaries, it is hoped that more holistic and fully integrated management may be possible. This approach to integrated coastal-zone management has great potential for developing sustainable management plans.

6.3.2   Heritage Coasts of England and Wales

Management schemes such as SCOPAC address obvious needs for coasts that are under population, development, tourist and industrial pressure. However, unspoilt and scenic coastlines require management as well, and this has led in England and Wales to the setting up of Heritage Coasts which aim 'to conserve the quality of the scenery [and] to foster leisure activities based on the natural scenery rather than man-made features' (Williams, 1992: 153). The Heritage Coasts scheme came about in 1973/74 following reports by the National Parks Commission (NPC), and is based on voluntary agreements between local councils and landowners on how best to manage a given coastline. It is of very low budget, with 50% of the financial contributions made by local councils, and a further 50% by the NPC (and its successors the Countryside Commission, the Countryside Agency (England), and the Countryside Council for Wales). The money is used to employ a Heritage Coast Officer and rangers, with some limited funds for building and protection work. At present, 1041 km of coastline is designated as Heritage Coast status in England, which is 32% of the entire English coastline (Table 6.2), and some 500 km in Wales. Many Heritage Coasts have been deemed successful in their aims (Williams, 1992), which demonstrates that non-

Table 6.2 Heritage Coasts of England

Coast	Defined (year)	Length (km)	County/region
Dover to Folkstone	1975	7	Kent
East Devon	1984	27	Devon
Exmoor	1991	45	Devon/Somerset
Flamborough Headland	1989	19	Yorkshire
Godrevy-Portreath	1986	9	Cornwall
Gribben Head-Polperro	1986	24	Cornwall
Hamstead	1988	11	Isle of Wight
Hartland	1986	11	Cornwall
Hartland	1990	37	Devon
Isles of Scilly	1974	64	Isles of Scilly
Lundy	1990	14	Devon
North Devon	1992	32	Devon
North Norfolk	1975	64	Norfolk
North Northumberland	1992	110	Northumberland
North Yorkshire and Cleveland	1981	57	North Yorkshire/ Cleveland
Pentire Point-Widemouth	1986	52	Cornwall
Penwith	1986	54	Cornwall
Purbeck	1981	50	Dorset
Rame Head	1986	8	Cornwall
St Agnes	1986	11	Cornwall
St Bees Head	1992	6	Cumbria
South Devon	1986	75	Devon
South Foreland	1975	7	Kent
Spurn	1988	18	Yorkshire
Suffolk	1979	57	Suffolk
Sussex	1973	13	Sussex
Tennyson	1988	34	Isle of Wight
The Lizard	1986	27	Cornwall
The Roseland	1986	53	Cornwall
Trevose Head	1986	4	Cornwall
West Dorset	1984	41	Dorset
Total		**1041**	

Source: Reproduced with permission from the Countryside Agency's website:
http://www.countryside.gov.uk/what/hcoast/heri_tbl.htm

statutory management schemes can work. Indeed, this success is inspiring other citizen-based coastal conservation activities elsewhere (e.g. Kawabe, 2004).

6.3.3 Coastal zoning – the Great Barrier Reef Marine Park

The zoning of the coast is a relatively simple and yet effective way of managing and separating incompatible uses. Activities in zones may be allowed, allowed with permission (i.e. to licence-holders) or forbidden, and can be applied in economic, development, tourist or conservation situations. The Great Barrier Reef Marine Park (GBRMP) has adopted such a zonation strategy. Because of its size the park has been divided up into sections, so that the individiual sections are more easily managed and human impacts reduced. Examples of the kinds of zones defined for the Cairns section of the GBRMP include zones for general use, habitat protection, conservation parks, buffers, national parks, and preservation (Fig. 6.1; Kay and Alder, 2005). Although it is agreed that such zoning is an excellent mechanism for managing the park on the broad scale, it requires additions at the local level. For example, in the Cairns section, the Green Island Management Plan was brought into effect in 1993 to formalise the use of the resources of Green Island, so as to permit recreational activity, but also to protect its natural assets (Fig. 6.2; Plate 6.8).

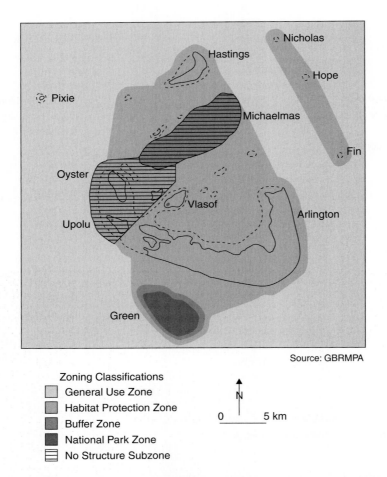

Source: GBRMPA

Zoning Classifications
- ☐ General Use Zone
- ☐ Habitat Protection Zone
- ■ Buffer Zone
- ■ National Park Zone
- ☰ No Structure Subzone

N

0 ___ 5 km

Figure 6.1 Coastal zoning approach to coastal management in the Cairns section of the Great Barrier Reef Marine Park (Australia).

Source: Kay and Alder (1999: 190, box 4.16).

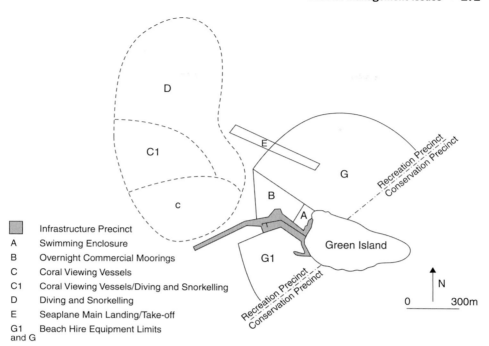

Figure 6.2 The zoning plan for Green Island in the Cairns section of the Great Barrier Reef Marine Park (Australia).

Source: Kay and Alder (1999: 125, box 4.4).

Plate 6.8 Information for tourists on the management zoning of Green Island in the Cairns section of the Great Barrier Reef Marine Park.

Technical Box 6.5

Remote sensing and geographic information systems – high-tech tools for the coastal manager

Managing the coast involves monitoring changes and the handling of much information. These tasks have been aided in recent years by the expansion of remote sensing techniques and geographic information systems (GIS). Remote sensing involves making observations some distance from the subject and, as far as the coastline is concerned, includes aerial photography, LIDAR (light detection and ranging) and satellite imagery, which can then be used in a GIS and/or a digital elevation or terrain model. This suite of techniques enables very long coasts to be surveyed and analysed quickly, providing comprehensive overviews. These techniques allow coasts to be mapped and changes monitored, such as ecosystem changes (e.g. Pasqualini *et al.*, 2001; Yang, 2005) and the geomorphology of barrier islands, beaches and dunes (e.g. Judge and Overton, 2001; Andrews *et al.*, 2002; Saye *et al.*, 2005). GIS techniques have even been applied to mapping Holocene coastal palaeogeography (e.g. Berendsen *et al.*, 2007).

These techniques not only allow physical changes to be measured, but also pollution and water quality to be detected and monitored (Liu *et al.*, 2003). They have been particularly useful in tracking the development of oil spills, such as for the Exxon Valdez spill in Alaska (Stringer *et al.*, 1992). A number of instruments have been introduced to monitor coasts, such as NASA's Coastal Zone Colour Scanner (CZCS). These data and information collected on the ground require storage, and GIS are rapidly becoming the main storage and retrieval system for such information. GIS allows layers of information to be superimposed, all of which can be interrogated by the user, and used to inform management decisions (O'Reagan, 1996). Used together, remote sensing and GIS techniques are powerful tools in managing coastal problems (e.g. Populus *et al.*, 1995).

Summary points

- Coastal management aims to facilitate sustainable use of the coastal zone and to protect human interests at the coast from natural and human-related processes.
- Coastal management is issue-led, and a number of key issues are identified, including pressure brought about through population growth, various uses of the coast, coastal hazards, and administrative issues.
- Coastal management operates at many spatial scales, but to be fully effective small authorities should be integrated to provide overall consistent management along a stretch of coastline.
- Legislation is often required, but some non-statutory management schemes have proved successful.

Discussion questions

1. Citing at least one example, assess the impact of coastal population growth on the coastal environment.

2. Examine and prioritise the range of issues that should be addressed by the coastal manager.
3. Evaluate the benefits of integrated coastal-zone management.

Further reading

See also

Definitions of the coastal zone, section 1.1.1, Management Box 1.1
Waves and coastal management, Management Box 2.1
Impact of tourism on coral cays, Case Study Box 2.4
Human interference in bar-built estuaries, Management Box 3.3
Deltas and human activity, section 4.8
Managing global sea-level rise, section 5.4.7

Introductory reading

Coastal Planning and Management. R. Kay and J. Alder. 2005. E & FN SPON, London, 380pp.
   A useful overview of planning and management strategies, provisioned with ample case studies from around the world.

Coastal and Estuarine Management. P. W. French. 1997. Routledge, London, 251pp.
   Examines the impact of different activities on the coast.

Coastal Defences: Processes, Problems and Solutions. P. W. French. 2001. Routledge, London, 366pp.
   Reviews and outlines engineering methods to defend eroding coasts.

Handbook of Coastal Engineering. J. B. Herbich (ed.). 2000. McGraw-Hill, New York, 1152pp.
   A very comprehensive volume of state-of-the-art management practices and methods.

Advanced reading

Coastal Management 2003. R. McInnes (ed.). 2003. Thomas Telford, London, 383pp.
   Collection of advance engineering research papers in the field of coastal engineering and management.

From beaches to beach environments: linking the ecology, human-use and management of beaches in Australia. R. J. James. 2000. *Ocean and Coastal Management,* **43,** 495–514.
   A useful overview of the varied factors that make up beach environments; also examines appropriate management systems.

The Geographical Journal, volume **164,** part 3. R. J. Nicholls and J. Branson (eds). 1998.
   A special issue on coastal management in northwest Europe.

Beach Management. E. C. F. Bird. 1996. Wiley, Chichester, 281pp.
   Examines the geomorphological and ecological aspects of beach management.

Studies in European Coastal Management. P. S. Jones, M. G. Healy and A. T. Williams. 1996. Samara Publishing Limited, Cardigan, 292pp.
   A very useful set of papers on coastal management in a European context.

Conclusion

The coast is a complex environment occupying a moveable and difficult-to-define space spanning the passage between land and sea, where a multitude of processes, driven by different energy sources and transferred through various agents, interact over time with the earth, and society.

This concluding statement probably sums up, in a single sentence, the content and approach of this book, which I began by exploring definitions and classifications of the coast, described a conceptual framework within which coastal systems may be understood, and signposted the importance of grasping the terminology and relatively extensive lexicon used by coastal scientists. Armed with this underpinning, the reader should be able to appreciate wave and tidal processes, and the coastal landforms and management issues genetically related to them. The decision to include a chapter on deltas (Chapter 4) was justified as perhaps the most salient example of how coastlines are influenced not only by oceanic and atmospheric processes but also by terrestrial factors. Also, given that large human populations inhabit deltas, it is important that we understand this fertile but often vulnerable landscape.

Many would argue that sea-level change is such a fundamental parameter for coastal evolution that a chapter on it should have appeared early on in the book, rather than as the penultimate chapter, as I have done here. However, many years of teaching the subject has made it clear to me that it is difficult for the student to appreciate the significance of sea-level change without first being introduced to the coastal processes and landforms that occur wherever the contemporary sea level determines them to be. Furthermore, a discussion on sea-level change leads logically into coastal management issues, as sea-level rise related to global warming is currently a major issue being considered by coastal managers. Readers will have realised that it is not the intention of this book to provide a guide to coastal management strategies and practices, such as the framework of integrated coastal-zone management (ICZM), but rather to exemplify the diversity of issues that need to be addressed by coastal managers.

If I were to write learning outcomes from this book they would include an appreciation of how coastal systems are constructed and operate, the diversity of coastal processes, landforms and landscapes, the realisation that coastal change occurs on varying scales of time, space and magnitude, and that humans affect, and are affected by, coasts. Perhaps on a higher academic level, I might also add the ability to critically evaluate theories and field evidence in the pursuit of a deeper understanding of coastal systems. I hope that a number of readers may indeed go on to attain this higher level and may even make original contributions to the field of coastal research.

At the beginning of the twenty-first century it has been suggested that the immediate future for coastal studies lies in the harnessing of new technology alongside specialist knowledge and numerical modelling to advance our understanding of coasts (Stephenson and Brander, 2003). I would agree with this, but would emphasise that technology is just a tool that can only be used to help answer questions that we ask based on the status of our knowledge about coastal systems. It surprises me how many researchers now write about coastlines they have not visited, instead relying on remotely sensed images and data, or model outputs. This perhaps shouldn't be surprising given the rise in digital technology, and such research is often, of course, extremely valuable. Nevertheless, after reading this book, I would urge students to (and hope they will if they can) adopt a field approach to coastal studies, at least initially, so that theoretical knowledge and empirical observations continue to enhance one another, and also to allow a personal understanding of coastal systems to develop and grow. Indeed, even remotely sensed images should be ground-truthed, and model output checked and verified through fieldwork.

Finally, we are regularly reminded of the very serious consequences of failing to appreciate and understand coastal systems. In the past, coastal disasters that have occurred have largely been classed as natural phenomena, but the landfall of Hurricanes Katrina and Rita in 2005 stimulated people to ask whether global warming was finally having an influence on climate and the meteorological conditions that gave rise to the storm surge that flooded New Orleans and the surrounding region. Whether or not there is a link can be debated, but the asking of the question indicates that the importance of understanding natural systems, and the consequences of interfering in them, has now escaped the confines of academia and is pervading society. In a similar way, in tragic circumstances, the word tsunami escaped academic circles in 2004 to enter common usage, but inspirationally a number of people said they survived the Indian Ocean event because they had learned about tsunami in school or college and were able to read the signs that one was approaching the coast. In addition to the interest and excitement the study of coasts offers students and researchers, it is absolutely clear that an understanding of coastal systems is important for the safety of all individuals who visit the coast, and the sustainability of coastal communities in the future. I hope this book has provided more than a basic understanding, and extended an invitation to pursue further study in the fascinating and worthwhile field of coastal systems.

Glossary

λ

Symbol for wave-length.

abrasion

The wearing down of rock through frictional contact with rock fragments.

accretion

The addition of newly deposited sediment.

advection

The physical mixing of fluids of different density (e.g. fresh and salt water in an estuary).

afro-trailing edge coast

Intermediate type of trailing edge coast.

allochthonous

Describes material (sediment) transported to the site of deposition.

amero-trailing edge coast

Mature type of trailing edge coast.

amino acid racemization

The alteration of L-isoleucine to D-alloisoleucine amino acids in shell material following death. The L/D ratio is used as a relative dating technique.

amphidromic motion, point and system

The circular motion of a tidal wave around a fixed point due to the Coriolis effect.

Atlantic type

A coast where rock structure is perpendicular to the coastline.

attrition

Reduction of rocks or rock fragments through erosional processes.

autochthonous

Describes material (sediment) deposited *in situ*.

backshore

A morphological term describing the part of a beach that lies between mean high water and the landward limit of marine (storm wave) activity.

backwash

Water returning to the sea under gravity following the breaking of a wave.

bar-built estuary

An estuary with a sedimentary feature, such as a bar or spit, partly obscuring the estuary mouth.

bay-head beach

An accumulation of sand that is trapped in an embayment by circulating currents, forming a beach.

beach cusp

Rhythmical longshore scallop-shaped features on a beach, comprising an upstanding horn and a depressed embayment.

beach zone
 A sedimentological term for the area between mean low water and the landward limit of marine (storm wave) activity.

beachrock
 A sand-sized deposit cemented by the precipitation of calcium carbonate. Occurs commonly in the tropics, but also in other areas, such as the Mediterranean.

bedload transport
 The movement of sediment along a surface near the bottom layer of the transporting fluid (air or water).

berm
 A ridge situated relatively high up on a beach marking the limit of normal swash action.

bio-erosion
 Erosion caused by the activity of biological organisms.

bird's-foot delta
 A river-dominated delta with semi-radiating channels separated by water that appear like a bird's footprint in aerial view.

black box system
 A system where the inputs and outputs are known, but the pathway through the system is not.

blind estuary
 An estuary whose connection to the sea is severed by the extension of a spit, for example, across the estuary mouth.

blowhole
 A vertical hole in a cliff-top connected to a sea-cave. Under certain conditions, water is forced out of the blowhole to produce a geyser.

blow-out
 An erosion hollow in coastal dunes formed by localised deflation, often in response to vegetation undermining by animals or people.

bottomset beds
 Horizontally layered sediment beds deposited in front of a delta as it pro-grades seaward. They become covered and end up at the bottom of a stack of delta sediments.

braided river
 A river comprising many interlaced channels.

braidplain
 A plain supporting more than one braided river.

breaker zone
 A process-based term for the area in the nearshore zone where incoming waves begin to break.

break-point
 The point at which a wave begins to break during shoaling.

Bruun rule
 A model suggesting the response of a sandy shoreline to sea-level change is to retreat landward.

buoyancy-dominated river mouth
 A deltaic river mouth where the river water is less dense than sea water and is, therefore, buoyant.

cascading system
 An approach describing the movement of energy and material through a system.

cay

An island (sometimes vegetated) situated on a coral reef constructed from eroded coral debris and concentrated by wave refraction.

circulatory cell

Refers to the circular pattern of currents and associated sediment movement in an enclosed coastal embayment.

clapotis

A standing wave that goes up and down, but does not appear to progress through the water.

closed system

A system where energy only transfers across the system boundaries.

coastal squeeze

The narrowing of the intertidal zone due to sea-level rise in conjunction with permanent barriers (human built or natural) preventing the landward migration of the intertidal zone.

coastal zone

The area between the seaward limit of terrestrial influence and the landward limit of marine influence.

constructive waves

These waves build up beach relief when long-period waves cause net landward sediment transport, most commonly during fairweather wave conditions.

Coriolis effect

An effect whereby the spinning of the earth makes travelling objects appear to move to the right in the northern hemisphere and to the left in the southern hemisphere.

corrasion

The mechanical weathering of rock surfaces by abrasion and other related processes.

coseismic subsidence

Subsidence of land coincident with an earthquake.

crevasse channel and splay

A channel formed by river water breaking through a levée and the splay of sediment deposited by such a breach.

cross-bedding

Bedding planes in sediment that are at oblique angles to one another. Most often present in dune deposits where bedding planes representing both stoss- and lee-sides may be present.

cuspate foreland

In plan view, a triangular-shaped depositional landform protruding seaward of the coastline, often arising where there is an abundant sediment supply and where two opposing longshore drift systems meet, piling up sediment into this example of an equilibrium trap. The point of the triangular cuspate foreland usually faces the direction of minimum fetch.

deep sink

Any store of sediment below the influence of waves and onshore currents. Sediment that finds its way into a deep sink is lost from the coastal system until reworked under different sea levels.

deflation

The erosion and removal of sand by aeolian processes.

destructive waves

Waves that lower beach relief under short-period wave conditions, usually associated with storm wave conditions.

diatoms

Unicellular algae that inhabit all aquatic environments. They possess a frustule composed of silica and are often preserved as microfossils in coastal sediments making them useful in palaeoenvironmental studies.

differential erosion

Uneven erosion of a coastline due to local variations in the properties (e.g. hardness, structure) of adjacent rock types.

diffraction

A process whereby waves can enter wave shadow zones (e.g. in the lee of an island or other obstacle) due to gravity allowing water to propagate or 'bleed' laterally out of the side of a wave 'severed' by the obstacle.

diffusion

The mixing of fluids of different ionic compositions in order to obtain uniformity (e.g. fresh and salt water in an estuary).

dispersion

The separation of different wave trains, through variations in wave travel time, i.e. long wave-length waves travel faster than short wave-length waves.

dissipative domain

A beach which dissipates and absorbs wave-energy.

distributary channels

River channels on a delta that commonly diverge as they approach the shore.

diurnal tide

A tidal cycle possesing one high and one low tide per day.

dominant current

Refers to the dominant current in a tidal environment, i.e. flood- or ebb-tide currents.

drift-aligned

Describes a coast that is aligned obliquely to prevailing incident waves.

dune slack

A depression in a coastal dune system that often separates dune ridges.

dynamic equilibrium

A state of equilibrium where environmental conditions fluctuate around an average that is itself gradually changing.

ebb tide

Outgoing tidal flow.

ecosystem

The interaction of organisms with the physical environment.

edge waves

Reflected waves that travel in a direction quasi-parallel to the shore and interact with incident waves.

embryo dune

A mound of wind-blown sand accumulating around the high-water level and colonised by pioneer dune vegetation. Embryo dunes often develop into foredunes.

endogenetic processes

Internal earth processes (e.g. plate tectonics) driven by geothermal energy.

equilibrium trap

A coastal sediment trap commonly occurring where two opposing longshore drift systems meet. Examples include cuspate forelands and tombolos.

estuary mouth bypassing

The offshore removal of estuarine/deltaic sediment by ebb-tidal currents.

eustasy

Absolute level of the sea surface and its fluctuations.

eutrophication
The stagnation of an aquatic ecosystem through oxygen depletion brought about by high nutrient levels that over-stimulate biological production.

exogenetic processes
Processes operating on the earths surface, driven by solar energy.

facies
The characteristics of a sediment that represents the environment of deposition.

fairweather wave
Long-period, low-height waves that usually characterise summer wave conditions.

fan deltas
Deltas characterised by high levels of fluvial energy and consequently coarse-grained sediment deposition (there are a number of subtypes explained in the main text).

feeder system
The general term for a fan delta's fluvial sediment source.

fetch
The distance over which waves travel.

fish-hook beaches
Beaches that are separated by headlands and become wider down-drift between the headlands.

fjord
A glaciated valley drowned by sea-level rise.

flocculation
The coagulation of clay particles of river origin upon encountering salt nucleii in an estuary.

flocs
A relatively large particle formed by the coagulation of relatively small individual clay particles during the process of flocculation.

flood tide
Incoming tidal flow.

flood-tide delta
A deltaic accumulation of sediment landward of the mouth of an estuary, deposited by dominant flood-tide currents.

fluid threshold velocity
Velocity of a fluid required to initiate movement of sediment of a given particle size.

forced waves
Waves that are actively being generated.

foredune
The most seaward ridge of a coastal dune system.

foreset beds
Seaward-sloping sediment beds deposited at the delta front.

foreshore
A morphological term for the area on a beach between mean low and high water.

free dunes
Sand dunes that are not colonised by vegetation and so are free to move in response to changing wind conditions.

free waves
Waves that have left the area in which they were generated.

friction-dominated river mouth
Shallow delta river mouths where friction with bed forces the emerging flow to spread laterally along the coast.

fringing reefs
Coral reefs that fringe the mainland or continental islands.

geoid
    The uneven surface of the sea, varying by up to 100 m, due to gravity differences in the earth.

Gilbert-type delta
    A delta comprising characteristic bottomset, foreset and topset beds.

glacio-eustasy
    Changes in absolute sea level brought about by the formation of continental ice, extracting oceanic water during an ice age.

glacio-isostasy
    Changes in land level brought about by the growth and decay of continental ice. The weight of ice pushes the land down, which then rebounds on deglaciation. This influences regional relative sea levels.

grey box system
    A system where input and outputs are known, along with some appreciation of the internal workings of the system.

groyne
    An artificial structure positioned across a beach at right-angles to the shore to interrupt and trap sediment being transported by longshore currents.

H
    Symbol for wave-height.

H_s
    Symbol for significant wave-height.

halocline
    Level in the water column where salinity abruptly changes, i.e. the contact between fresh surface and deep saline waters in an estuary.

halophytes
    Plants able to tolerate saline conditions.

Holocene Epoch
    From 10 ka to the present day (current interglacial).

homogeneous estuary
    An estuary with uniform salinity throughout.

homopycnal flow
    Flow occurring where river water and the receiving water body possess equal densities.

hydraulic duty
    The volume of water transferred on or off a surface, e.g. a salt marsh surface.

hydraulic pressure
    Pressure created in a rock void through the trapping of air or water by breaking waves, serving to weaken rock.

hydrological cycle
    The transfer of water between natural storages.

hyperpycnal flow
    Flow occurring where river water is received by a water body of lower density.

hypersynchronous estuary
    An estuary where tidal range increases up-estuary.

hypopycnal flow
    Flow occurring where river water enters a water body of higher density.

hyposynchronous estuary
    An estuary where tidal range decreases up-estuary.

impact threshold velocity
    Velocity required to move sediment grains of a given particle size through the combined influence of the fluid velocity and energy transferred to the sediment grain by collision with other grains already in motion.

impeded dunes
Sand dunes that are semi-fixed by vegetation colonisation.

incident wave
A wave approaching a shore, usually for the first time, i.e. not a reflected wave.

indicative meaning and range
The position of a sea-level index point in the tidal frame at the time of formation or deposition.

inertia-dominated river mouth
River outflow in a delta setting that is unconfined, spreading out in all directions to form a lobate delta.

infauna
Animals living within sediment.

inner-shelf delta
A delta situated in shallow water in an inner shelf position.

inshore
A morphological term for the area between wave-base and mean low water.

interdistributary areas
Deltaic areas lying between distributary channels.

interseismic strain accumulation
Strain accumulated along a fault between earthquake events, often resulting in minor uplift.

intertidal zone
The area lying between high- and low-tide levels.

isolation basin
A coastal basin cut off and isolated from the sea by relative sea-level fall.

isostasy
The vertical movement of the land due to the loading and unloading of weights (e.g. ice).

karst
Features associated with the chemical dissolution of limestone, such as cave systems.

lagoonal estuary
An estuary whose connection to the sea is severed by the extension of a spit, for example, across the estuary mouth.

laterally homogeneous estuary
An estuary where salinity is uniform across, but changes along, the estuary.

laterally inhomogeneous estuary
An estuary where salinity changes across and along the estuary, but is uniform at all depths at any given point.

leading edge coast
A coast occurring on a continental margin close to a converging crustal plate boundary.

level of no motion
A level in the water column of an estuary separating opposing upper and lower water layers where no tidal flow occurs.

lithology
Rock type.

littoral energy fence
Onshore wave-energy that prevents sediment escaping offshore.

longshore drift
Currents that move along a shore, often transporting sediment.

luminescence dating
A method of dating minerals based on trapped luminescent energy since burial. Exposure to light bleaches the minerals and resets the luminescent clock back to zero.

lutocline
The gradient of suspended sediment concentration in the water column of an estuary.

macrotidal
A tidal environment with a tidal range >4 m.

managed retreat
Allowing and/or managing the landward retreat of a shoreline due to coastal erosion and/or sea-level rise.

marginal sea coast
A coast on a continental margin some distance from a converging crustal plate boundary.

mesotidal
A tidal environment with a tidal range of 2–4 m.

meta-stable equilibrium
A system that switches or jumps from one level of equilibrium to another due to trigger events.

microtidal
A tidal environment with a tidal range <2 m.

mixed tide
A tidal cycle comprising one dominant high and low tide per day, but with additional secondary (significantly smaller) high and low tides.

morphodynamic
Changing morphology in response to variations in environmental conditions.

morphological system
An approach that deals with the morphological expression of a system.

neap tide
Tide that occurs when the sun and moon are perpendicular with regard to the earth, to produce lower than average high tides and higher than average low tides.

nearshore zone
Process-based term for the area comprising swash, surf and breaker zones.

negative feedback
A response to a system change that aids recovery of the original state of equilibrium.

neo-trailing edge coast
Immature type of trailing edge coast.

non-turbid estuary
An estuary with sediment too coarse for it to be held in suspension for long periods of time, so that the water is relatively clear.

null point
The point on an estuary bed where opposing flows meet, reducing flow velocity to zero.

offshore
A morphological term for the area below wave-base.

offshore zone
A sedimentological term for the area lying below storm wave-base.

offshore transition zone
The area between fairweather and storm wave-bases.

open system
A system where energy and matter are transferred across its boundaries.

open-ended estuary
An estuary with an unobscured connection to the sea.

orthogonals
Lines drawn at right-angles to a wave-front and extended to the shore, indicating wave-energy concentration received along any given length of the coast, i.e. where

orthogonals converge wave-energy is concentrated, but is dissipated where orthogonals diverge.

oscillatory wave
A deep-water wave in which water particles move in an orbital motion.

Pacific type
A coast where the rock structure is parallel to the coastline.

parabolic dune
A crescentic impeded dune commonly found on the downwind perimeter of blow-outs, with its arms opening upwind.

parallel orientated bedforms
Sedimentary bedforms aligned parallel to the prevailing current direction.

partially closed estuary
An estuary with a sedimentary feature, such as a bar or spit, partly obscuring the estuary mouth.

particle size
The size of either individual sediment particles or entire populations.

patch reef
Subrounded platform-like coral reefs located on the continental shelf.

permeability
The ability of a rock type to allow water to pass through it via structural and bedding planes.

perpendicular orientated bedforms
Sedimentary bedforms aligned perpendicular to the prevailing current direction.

Pleistocene Epoch
The first epoch of the Quaternary, from 1.81 Ma to 10 ka.

plunging breakers
Breaking waves that characterise steeply inclined shores and comprise a steep-fronted wave that curls over, crashing down on the shore producing foam.

plunging cliffs
Quasi-vertical cliffs that extend below sea level. Commonly found on previously glaciated fjordic coasts where U-shaped valleys have been drowned and the steep sides forming plunging cliffs.

pocket beach
A small beach located between headlands.

positive feedback
A response to a system change that enhances the change leading to a new state of equilibrium.

pressure-realease jointing
Cliff-parallel joints created by the removal of rock from a cliff front, releasing confining pressure.

primary dune
Coastal sand dunes which are orientated parallel to the shore in their original depositional position.

process-response system
An approach which recognises that morphology is determined through response to processes.

progradation
The seaward advance of a sedimentary shoreline, usually through excess sediment input over output.

quarrying
The detachment and removal of rock fragments from cliffs and shore platforms by wave activity.

Quaternary
The most recent geological period from 1.81 million years ago to the present. Comprises the Pleistocene and Holocene Epochs and is characterised by alternating glacial and interglacial conditions.

radiocarbon dating
A method of dating carbonaceous material using the decay and half-life of ^{14}C.

raised beach
A fossil beach that occurs above present-day sea level.

re-entrant trap
A sediment trap, such as a bayhead beach or pocket beach, where sediment introduced into an embayment is unable to escape due to a circulatory cell operating within the embayment.

reflective domain
A steeply inclined beach state that reflects wave-energy back out to sea.

refraction
The bending of a wave front caused by a velocity reduction when the wave enters shallow water, so that the sea-floor is equal to or shallower than wave-base.

regression
Sea-level fall and the emergence of coastal land.

relative sea-level change
The level of the sea relative to the land, determined by eustasy and isostasy.

relief
Character of the land surface, including topography and elevation. High relief refers to mountainous or hilly terrain, whilst low relief refers to low and rather featureless terrain.

retrogradation
The landward retreat of a sedimentary shoreline, usually through a deficit of sediment input over output.

ria
A river valley drowned by sea-level rise.

ribbon reef
A linear coral reef orientated parallel to the shore and usually located along the seaward edge of the continental shelf.

rip current
A seaward-directed current associated with water returning to the sea after being brought onshore by wave-breaking activity.

river mouth bypassing
Sediment flushed out to sea by high river discharge (e.g. flood event).

rollover
The gradual net landward migration of a subaerial sediment barrier (mainly refers to gravel barriers).

runnels
Shore-parallel gullies occurring on the low-angled facet of a beach near the low-water level and separated by ridges, although may be connected through the ridges by rip channels.

sabka
Intertidal to supratidal salt flats that occur along arid coastlines, often characterised by algal mats, salt crusts and other evaporite minerals.

salient trap
A sediment trap that extends outwards from the shore, such as a spit.

salinisation
The increase of salt levels in soil due to high evaporation rates or saline groundwater.

salt pan
A pool of saline water found on a salt marsh surface.

saltation
A mode of sediment transport that involves sediment grains moving in a series of jumps or hops.

salt-balance principle
A model proposed by Pritchard (1955) to explain the mixing of fresh and salt water in an estuary, through the processes of advection and diffusion, with implications for the classification of estuaries.

salt-wedge
A distinct wedge of dense saline water that occurs at the bottom of an estuarine water column and tapers upstream. Common in estuaries where fresh/salt water mixing is limited.

sea-level index points
Points used to construct sea-level curves for which age, altitude, indicative meaning and range, and tendency of sea level are known.

sea waves
Wind-generated waves encountered within the generation area, often high and with short period.

secondary dune
A coastal dune that has been reworked by coastal processes to form a ridge at right-angles to the shore.

sediment couplet
Alternating coarse and fine sediment layers, often associated with tidal environments where the coarse sediment is deposited by the dominant current and the fine sediment by the subordinate current.

sediment sorting
Similarity of sediment populations in terms of size, shape or both. A sediment comprising grains of a similar size and/or shape is said to be well-sorted.

seiche
A type of standing wave, created by incident wave reflection, that sloshes about in enclosed coastal waters. Severe seiches in harbours may lift boats out of the water.

semi-diurnal tide
A tidal cycle comprising two high and two low tides per day.

settling lag
The phenomenon whereby sediment particles can undergo net shoreward transport in a tidal environment.

settling velocity
The velocity at which a sediment particle may settle out of the water column.

shadow dune
An unvegetated accumulation of wind-blown sand in the lee of an obstacle, often at the strandline where marine debris is plentiful.

shelf-edge delta
A delta built up at a site on the edge of a continental shelf through abundant sediment deposition outpacing the rise of sea level.

shoaling
The process that occurs when the wave-base of a deep-water oscillatory wave first intersects the sea-floor, so beginning the transition to a translatory wave, involving the formation of breaking waves.

shore platform
An eroded surface in the intertidal zone, often referred to as wave-cut, but other processes such as salt weathering and bio-erosion are also important.

shoreface bypassing
Offshore removal of coastal sediment by storm-induced offshore bottom currents.

shoreface zone
A sedimentological term for the area between mean low water and fairweather wave-base.

significant wave-height (H_s)
The mean of the highest one-third of waves encountered on a coast.

spilling breaker
A type of breaking wave that is characteristic of gently sloping beaches, producing lots of foam.

spit
A depositional landform that occurs where the shore orientation changes, but long-shore currents do not deviate and continue to transport and deposit along a projected coastline.

spring tide
A tide that occurs when the sun and moon are aligned with regard to the earth, to produce higher than average high tides and lower than average low tides.

standing wave
A wave that goes up and down, but does not appear to progress.

steady-state equilibrium
The situation where environmental conditions vary around an unchanging average state.

steric sea-level change
Sea-level change due to ocean water density changes, e.g. warm water is less dense, expands causing to sea level to rise.

stillstand
A period of time over which sea level does not change.

stillwater level
Mean water level, half-way between the depth of a wave trough and the height of the crest.

storm profile
A beach profile developed under storm wave conditions where destructive waves lower the beach gradient.

storm surge
Coastal inundation of sea water associated with storm conditions, especially high onshore winds that pile water up at the coast, and the effect of low pressure that allows sea level to rise by 1 cm for every millibar of pressure change.

submerged forest
Woodland that has been drowned by sea-level rise, preserved in peat, and subsequently exposed in intertidal and subtidal zones.

subordinate current
The least effective current in a tidal environment, i.e. flood- or ebb-tide currents.

summer profile
A beach profile developed under fairweather wave conditions where constructive waves build up the beach gradient.

superposition
The merging of two or more wave trains travelling in the same direction. Where the different waves are in phase (i.e. the crest and troughs of each wave train are superimposed) the wave amplitude is increased, whereas amplitude is decreased where the wave trains are not in phase. Gives rise to the phenomenon known as surf beat.

surf beat

The alternating arrival of sets of higher and lower than average wave amplitudes on the coast. See superposition.

surf zone

A process-based term for the area of the nearshore zone where breaking waves approach the shore, usually over a wide low-gradient surface.

surface creep

A mode of sediment transport which involves sediment grains being moved along a surface.

surging breaker

A type of breaking wave that surges up a steep beach by forcing water out of the front of the wave, producing little foam.

suspended particulate matter

Particles held in suspension in an estuary.

swash

The movement of water up the beach upon a wave breaking.

swash zone

Process-based term for the area of the nearshore zone where waves finally break and travel up the beach as swash. Also, water returns to the sea as backwash within this zone.

swash-aligned

The orientation of a shoreline parallel to approaching wave-fronts.

swell profile

A beach profile developed under fairweather wave conditions where constructive waves build up the beach gradient.

swell waves

Wind-generated waves encountered outside the area of generation, often of low height and long period.

synchronous estuary

An estuary where tidal range is uniform throughout.

T

Symbol for wave-period.

tendency of sea level

The direction of sea-level change indicated by a sea-level index point. Positive/ negative tendency refers to sea-level rise/fall.

tidal bore

A wave produced in a hypersynchronous estuary at the onset of the flood tide, where the estuary narrows, shallows and the tidal range increases upstream.

tidal bulge

Bulge of water pulled up by the gravitational attraction of the moon and the sun.

tidal creek

Creek which facilitates the flooding and withdrawal of tidal waters on and off a mudflat, salt marsh or mangrove surface.

tidal currents

Currents produced by the incoming and outgoing of tidal water at the coast, particularly effective in constricted waterways such as estuaries.

tidal cycle

The rise and fall of the tide as predicted by gravitational effects of the moon and sun on the sea surface.

tidal prism

The amount of water that enters and exits an estuary every flood and ebb tide, respectively.

tidal range
The vertical distance between high and low tide at a given location along the coast.

tidal rhythmites
Sediment couplets that reflect the rhythm of the tides, e.g. thicker sediment layers during spring tides and thinner during neap tides.

tide-dominated estuary
An estuary whose processes are dominated by tides.

tombolo
An accumulation of sediment extending from the shore to an offshore island or sandbank.

toppling failure
A mechanism of cliff erosion and retreat whereby a stack of rock topples forward, with the base of the stack remaining approximately in its original position, like a tree being felled.

topset beds
Horizontal sediment beds deposited on a delta plain.

trailing edge coast
A coast formed by crustal divergence with subsequent drift away from the plate boundary; it is tectonically stable.

transgression
The rise of sea level and coastal indundation.

translatory wave
A shallow-water wave in which water particles move in a highly elliptical motion.

tsunami
Waves created by submarine earthquakes, slides, volcanic activity and meteor/comet impacts in the ocean. Often tsunami can be catastrophic, particularly around the Pacific rim, and may make significant contributions to the development of coastlines in these areas.

turbid estuary
An estuary with sediment fine enough to be held in suspension for long periods of time, so that the water appears muddy.

turbidity
Refers to the amount of sediment held in suspension by a fluid. Clear water is non-turbid, whilst visibly sediment-rich fluid has high turbidity.

turbidity maximum
A zone in the waters of an estuary where maximum sediment concentration occurs.

uranium-series dating
A dating technique for mineral material that uses the decay and half-life of uranium, thorium, radon and lead.

wave-base
Equates to a water depth below which passing waves have no effect (variably defined as being equal to between a half and a quarter of the wave-length).

wave-cut notch
An indentation at the base of a cliff excavation by wave processes. Often leads to undermining of a cliff and subsequent cliff collapse.

wave-dominated estuary
An estuary whose processes are dominated by wave activity.

wave-height (H)
The vertical distance between the wave trough and crest.

wave-length (λ)
The horizontal distance between consecutive wave crests or troughs.

wave-period (T)

The time interval between consecutive wave crests or troughs passing a fixed point.

wet slack

A depression between coastal dune ridges where the surface intersects the water table creating wet conditions.

white box system

A system for which the inputs, ouputs and detailed workings are known.

winter profile

A beach profile developed under storm wave conditions where destructive waves lower the beach gradient.

xerophytic

Describes a plant able to withstand harsh environmental conditions, such as drought. Many plant species growing on coastal sand dunes, where water is scarce, are xerophytic.

zeta-form beach

Beaches that are separated by headlands and become wider down-drift between the headlands.

Further reading

General

Coasts: An Introduction to Coastal Geomorphology (3rd edn). E. C. F. Bird. 1984. Blackwells, Oxford, 320pp.
A standard introduction to coastlines.

Coastal Environments: An Introduction to the Physical, Ecological and Cultural Systems of Coastlines. R. W. G. Carter. 1988. Academic Press, London, 617pp.
A thorough advanced-level overview of the coast, quite technical in places, and becoming out-of-date.

The Evolving Coast. R. A. Davis, Jr. 1997. Scientific American Library, New York, 231pp.
A concise introduction to the physical development of coastlines.

Coastal and Estuarine Management. P. W. French. 1997. Routledge, London, 251pp.
A timely introduction to various management issues currently affecting coastal systems.

Coasts. J. D. Hansom. 1988. Cambridge University Press, Cambridge, 96pp.
A concise and readable introduction to coasts.

Les littoraux: Impact des aménagements sur leur évolution. R. Paskoff. 1998. Armand Collin, Paris, 260pp.
An excellent French language text dealing with coastal problems.

Introduction to Coastal Geomrophology. J. Pethick. 1984. Arnold, London, 260pp.
Although titled 'Introduction' this is in fact a fairly technical, and now slightly out-dated, guide to coastal geomorphology, but highly recommended for Pethick's elegant writing and excellent diagrams.

Exploring Ocean Science (2nd edn). K. Stowe. 1996. Wiley, New York, 426pp.
A very well illustrated and thorough review of oceanography. Only some of the chapters deal with the coastal ocean, but it covers processes, landforms and ecology.

Coastal Dynamics and Landforms. A. S. Trenhaile. 1997. Oxford University Press, Oxford, 365pp.
An advanced-level text dealing with process and form at the coast.

Coastal Problems: Geomorphology, Ecology and Society at the Coast. H. Viles and T. Spencer. 1995. Arnold, London, 350pp.
A readable text on problems affecting the global coastline.

Coastal Geomorphology: An Introduction. E. Bird. 2000. Wiley, Chichester, 322pp.
An extensive overview of coastal processes and geomorphology.

Coasts: Form, Process and Evolution. C. D. Woodroffe. 2002. Cambridge University Press, Cambridge, 623pp.
An encyclopaedic treatment of coasts for advanced undergraduate and graduate students.

Introduction to Coastal Processes and Geomorphology. G. Masselink and M. G. Hughes. 2003. Arnold, London, 354pp.
An accessible advanced introduction to coastal processes and geomorphology.

Dynamics of coastal systems. J. Dronkers. 2005. World Scientific, Singapore, 519pp.
An advanced text in ocean engineering aspects of coastal systems.

The periodicals *Journal of Coastal Research, Estuarine, Coastal and Shelf Science, Shore and Beach, Journal of Coastal Conservation, Marine Geology* and others should be consulted regularly for individual research papers, themed sections, and special issues devoted to particular topics.

Specific topics

Waves, tides and currents

Waves, Tides and Shallow-water Processes. Open University. 1989. Pergamon Press, Oxford, 187pp.
A thorough text focusing on the operation of coastal processes.

Waves in Oceanic and Coastal Waters. L. H. Holthuijsen. 2007. Cambridge University Press, Cambridge, 404 pp.
Relatively advanced text covering wind waves in detail.

Tides, Surges and Mean Sea-Level. D. T. Pugh. 1987. Wiley, Chichester, 472pp.

Tsunamis in the World. S. Tinti (ed.). 1992. Kluwer Academic, Dordrecht, 228pp.

Tsunami: The Underrated Hazard (2nd edn). E. Bryant. 2008. Praxis Publishing Ltd, Chichester, 330pp.
An accessible text questioning the role of tsunami on coastal evolution and examining their geomorphological significance.

Rocky coasts

The Biology of Rocky Shores. C. Little and J. A. Kitching. 1996. Oxford University Press, Oxford, 240pp.

Geomorphology of Rocky Coasts. T. Sunamura. 1992. Wiley, Chichester, 302pp.
An excellent and technical account of erosional processes and resultant landforms at the coast.

The Geomorphology of Rock Coasts. A. S. Trenhaile. 1987. Oxford University Press, Oxford, 384pp.
A thorough advanced-level overview of processes and landforms of erosional coastlines.

Rock coasts, with particular emphasis on shore platforms. A. S. Trenhaile. 2002. *Geomorphology*, **48**, 7–22.
A timely review of the processes and landforms of rocky coastlines.

Intertidal Ecology. D. Raffaelli and S. Hawkins. 1996. Chapman and Hall, London, 356pp.
A timely general introduction to the ecology of intertidal environments along all types of coasts.

Coral reefs

World Atlas of Coral Reefs. M. Spalding, C. Ravilious and E. P. Green. 2001. University of California Press, Berkeley, 424pp.
A colourful introduction to the biogeography of coral reefs around the world.

Oceanographic Processes of Coral Reefs: Physical and Biological Links in the Great Barrier Reef. E. Wolanski (ed.). 2001. CRC Press, London and Boca Raton. 356 pp.

A collection of papers on the physical, biological and management aspects of the Great Barrier Reef – useful case study material.

Coral Reef Geomorphology. A. Guilcher. 1988. Wiley, Chichester, 228pp.
An introduction to the physiology of coral reef systems throughout the world.

The Geomorphology of the Great Barrier Reef: Quaternary Development of Coral Reefs. D. Hopley. 1982. Wiley, New York, 453pp.
An exhaustive account of one of the world's most spectacular coastal environments.

Discover the Great Barrier Reef Marine Park (rev edn). L. Murdoch (compiler). 1996. Angus & Robertson, Sydney, 96pp.
A very well-illustrated and accessible introduction to the physical setting, ecology, history, conservation and management of Australia's Great Barrier Reef.

Ecosystems of the World (volume 25): Coral Reefs. Z. Dubinsky (ed.). 1990. Elsevier, Amsterdam.

Oceanic Islands. P. D. Nunn. 1994. Blackwell, Oxford.

Beaches, barrier islands and coastal dunes

Beaches: Form and Process. J. Hardisty. 1990. Unwin Hyman, London, 324pp.
An advanced-level text introducing beach and nearshore dynamics.

Beach and nearshore sediment transport. J. Hardisty. 1994. In *Sediment Transport and Depositional Processes* K. Pye (ed.). Blackwell Scientific Publications, Oxford, 219–255.

Beaches and Coasts. R. A. Davis Jr and D. M. Fitzgerald. 2004. Blackwell, Oxford, 419pp.
A comprehensive treatment of coastal geology and geomorphology.

Coastal Dunes: Form and Process. K. F. Nordstrom, N. Psuty and B. Carter (eds). 1990. Wiley, Chichester, 392pp.
A collection of papers dealing with the development and geomorphology of coastal dunes with good case study material.

Coastal Dune Management. D. Ranwell and R. Boar. 1986. Institute of Terrestrial Ecology, Huntingdon, 105pp.

Coastal Dunes: Ecology and Conservation. M. L. Martínez and N. P. Psuty (eds). 2004. Springer, Berlin, 390 pp.
A collection of chapters from a number of well-respected authors in this field covering a wide range of dune topics from geomorphology to conservation.

Marine Biology: Function, Biodiversity and Ecology. J. S. Levington. 1995. Oxford University Press, Oxford.
Recommended for its treatment of beach ecosystems.

Tidal environments

Estuaries: A Physical Introduction (2nd edn). K. R. Dyer. 1997. Wiley, Chichester, 210pp.
An introductory text to estuaries, including their classification, tidal effects and mixing processes, but quite mathematical in places.

Estuarine Shores: Evolution, Environments and Human Alterations. K. F. Nordstrom and C. T. Roman (eds). 1996. Wiley, New York, 510pp.
An advanced compilation of work on the very varied nature of estuarine environments.

The Estuarine Ecosystem (2nd edn). D. S. McLusky. 1989. Blackie, Glasgow, 215pp.
A now standard and well-accepted introduction to estuarine ecosystems.

Saltmarshes: Morphodynamics, Conservation and Engineering Significance. J. R. L. Allen and K. Pye (eds). 1992. Cambridge University Press, Cambridge.
A useful and wide-ranging introduction to salt marsh systems, their ecology and applied issues.

Intertidal Ecology. D. Raffaelli and S. Hawkins. 1996. Chapman and Hall, London, 356pp.
A timely general introduction to the ecology of intertidal environments along all types of coasts.

Mangrove – The Forgotten Habitat. J. S. Deitch. 1996. Immel Publishing, London.

The Botany of Mangroves. P. B. Tomlinson. 1986. Cambridge University Press, Cambridge.

High Resolution Morphodynamics and Sedimentary Evolution of Estuaries. D. M. Fitzgerald and J. Knight (eds). 2005. Springer, Berlin, 364 pp.
A collection of papers that examine the relationships between sediments and estuarine landforms.

Estuaries: Monitoring and Modelling the Physical System. J. Hardisty. 2007. Blackwell, Oxford, 157pp.
A useful instruction manual to creating an estuarine model, with the Humber Estuary, UK, used as an example.

Coastal and Estuarine Environments: Sedimentology, Geomorphology and Geoarchaeology. K. Pye and J. R. L. Allen (eds). 2000. Geological Society, London, Special Publication No. 175, 470pp.
A collection of advanced research papers on sedimentary, and especially estuarine, coasts.

Academic journals, such as *Estuaries, Estuaries and Coasts,* and *Estuarine, Coastal and Shelf Science,* are worth consulting regularly for current science in this area.

Deltaic environments

Clastic coasts. H. G. Reading and J. D. Collinson. 1996. *In:* H. G. Reading (ed.) *Sedimentary Environments: Processes, Facies and Stratigraphy.* Blackwell, Oxford, 154–231.
A substantial chapter on modern and ancient clastic coasts, but with a clear emphasis on deltaic sedimentary environments.

Journal of Coastal Research, volume **14**, part 3. 1998.
A timely thematic issue devoted to deltaic coastal environments, with some excellent review articles, including the Mississippi Delta and Arctic deltas, as well as research papers. A good source of recent delta literature.

Coarse-grained Deltas. A. Colella and D. B. Prior (eds). 1990. Blackwell Scientific Publications, Oxford, 357pp.
A compilation of research papers on fan deltas.

Deltas: Sites and Traps for Fossil Fuels. M. K. G. Whateley and K. T. Pickering (eds). 1989. Geological Society, London, Special Publication No. 41, 360pp.
As the title suggests, this is a collection of geological research papers devoted to hydrocarbon exploration in deltaic regions.

Deltaic and estuarine environments. P. French. 2007. *In:* C. Perry and K. Taylor (eds) *Environmental Sedimentology.* Blackwell Publishing, Oxford, 223–262.
A wide-ranging yet brief overview of deltaic sedimentary environments.

Records of sea-level change

Sea-Level Changes: The Last 20 000 Years. P. A. Pirazzoli. 1996. Wiley, Chichester, 211pp.
A well-written introduction to sea-level changes on a geological timescale.

Ice Age Earth: Late Quaternary Geology and Climate. A. G. Dawson. 1992. Routledge, 293pp.
An extensive review of climate and environmental change from the last glacial onwards.

Sea Level Rise: History and Consequences. B. C. Douglas, M. S. Kearney and S. P. Leatherman. 2000. Academic Press, New York, 232pp.

Examines historic sea-level rise and is accompanied by a CD-Rom containing historic sea-level data.

Sea-level Research: A Manual for the Collection and Evaluation of Data. O. van de Plassche (ed.). 1986. Geo Books, Norwich, 618pp.

The standard reference work for the collection and interpretation of sea-level data.

Quaternary Environmental Micropalaeontology. S. K. Haslett (ed.). 2002. Arnold, London, 340pp.

A thorough overview of the use of microfossils in the study of Quaternary palaeoenvironments, including sea-level research applications.

Impacts of sea-level rise

Climate Change 2007: The Physical Science Basis. Contribution of Working Group I to the Fourth Assessment Report of the Intergovernmental Panel on Climate Change. Solomon, S., D. Qin, M. Manning, Z. Chen, M. Marquis, K.B. Averyt, M. Tignor and H.L. Miller (eds). 2007. Cambridge University Press, Cambridge.

Latest IPCC science report.

Climate Change 2007: Impacts, Adaptation and Vulnerability. Contribution of Working Group II to the Fourth Assessment Report of the Intergovernmental Panel on Climate Change. M.L. Parry, O.F. Canziani, J.P. Palutikof, P.J. van der Linden and C.E. Hanson (eds). 2007. Cambridge University Press, Cambridge.

Latest IPCC impact report.

Submerging Coasts: The Effects of a Rising Sea Level on Coastal Environments. E. C. F. Bird. 1993. Wiley, Chichester, 184pp.

A useful review of the response of coastlines to rising sea levels.

Climate and Sea-level Change: Observations, Projections, and Implications. R. A. Warrick, E. M. Barron and T. M. L. Wigley (eds). 1993. Cambridge University Press, Cambridge, 424pp.

A collection of papers examining the issue of global climate change and associated sea-level rise.

Impacts of Sea-level Rise on European Coastal Lowlands. S. Jelgersma and M. J. Tooley (eds). 1992. Blackwell, Oxford, 267pp.

An example of how sea-level rise impacts on a regional scale.

Present-day sea-level change: a review. R. S. Nerem, E. Leuliette and A. Cazenave. 2006. *Comptes Rendus Geoscience*, **338**, 1077–1083.

A brief review of observed sea-level change for the last 50 years.

Coastal management

Coastal Planning and Management (2nd edn). R. Kay and J. Alder. 2005. E & FN SPON, London, 380pp.

A useful overview of planning and management strategies, provisioned with ample case studies from around the world.

Coastal and Estuarine Management. P. W. French. 1997. Routledge, London, 251pp.

Examines the impact of different activities on the coast.

The Geographical Journal, volume **164**, part 3. R. J. Nicholls and J. Branson (eds). 1998.

A special issue on coastal management in northwest Europe.

Studies in European Coastal Management. P. S. Jones, M. G. Healy and A. T. Williams. 1996. Samara Publishing Limited, Cardigan, 292pp.

A very useful set of papers on coastal management in a European context.

Directions in European Coastal Management. M. G. Healy and J. P. Doody. 1995. Samara Publishing Limited, Cardigan, 556pp.
The conference proceedings of Coastlines '95, contains a number of very useful papers.

Coastal Defences: Processes, Problems and Solutions. P. W. French. 2001. Routledge, London, 366pp.
Reviews and outlines engineering methods to defend eroding coasts.

Handbook of Coastal Engineering. J. B. Herbich (ed.). 2000. McGraw-Hill, New York, 1152pp.
A very comprehensive volume of state-of-the-art management practices and methods.

Beach Management. E. C. F. Bird. 1996. Wiley, Chichester, 281pp.
Examines the geomorphological and ecological aspects of beach management.

Bibliography

Allen, J. R. L., 1989. Evolution of salt-marsh cliffs in muddy and sandy systems: a qualitative comparison of British west-coast estuaries. *Earth Surface Processes and Landforms*, **14**, 85–92.

Allen, J. R. L., 1993. Muddy alluvial coasts of Britain: field criteria for shoreline position and movement in the recent past. *Proceedings of the Geologists' Association*, **104**, 241–262.

Allen, J. R. L., 1994. A continuity-based sedimentological model for temperate-zone tidal salt marshes. *Journal of the Geological Society, London*, **151**, 41–49.

Allen, J. R. L., 1996. Shoreline movement and vertical textural patterns in salt masrh deposits: implications of a simple model for flow and sedimentation over tidal marshes. *Proceedings of the Geologists' Association*, **107**, 15–23.

Allen, J. R. L., 1999. Geological impacts on coastal wetland landscapes: some general effects of sediment autocompaction in the Holocene of northwest Europe. *The Holocene*, **9**, 1–12.

Allen, J. R. L., 2000a. Morphodynamics of Holocene salt marshes: a review sketch from the Atlantic and Southern North Sea coasts of Europe. *Quaternary Science Reviews*, **19**, 1155–1231, 1839–1840 (erratum).

Allen, J. R. L., 2000b. Holocene coastal lowlands in NW Europe: autocompaction and the uncertain ground. *In:* K. Pye and J. R. L. Allen (eds) *Coastal and Estuarine Environments: Sedimentology, Geomorphology and Geoarchaeology*. Geological Society, London, Special Publication No. 175, 239–252.

Allen, J. R. L., 2003. An eclectic morphostratigraphic model for the sedimentary response to Holocene sea-level rise in northwest Europe. *Sedimentary Geology*, **161**, 31–54.

Allen, J. R. L. and Haslett, S. K., 2002. Buried salt-marsh edges and tidal-level cycles in the mid Holocene of the Caldicot Level (Gwent), South Wales. *The Holocene*, **12**, 303–324.

Allen, J. R. L., and Haslett, S. K., 2006. Granulometric characterization and evaluation of annually banded mid-Holocene estuarine silts, Welsh Severn Estuary (UK): coastal change, sea level and climate. *Quaternary Science Reviews*, **25**, 1418–1446.

Allen, J. R. L. and Haslett, S. K., 2007. The Holocene estuarine sequence at Redwick, Welsh Severn Estuary Levels, UK: the character and role of silts. *Proceedings of the Geologists' Association*, **118**, 157–185.

Allen, J. R. L., Haslett, S. K. and Rinkel, B. E., 2006. Holocene tidal palaeochannels, Severn Estuary Levels, UK: a search for granulometric and foraminiferal criteria. *Proceedings of the Geologists' Association*, **117**, 329–344.

Alve, E. 1995. Benthic foraminiferal responses to estuarine pollution: a review. *Journal of Foraminiferal Research*, **25**, 190–203.

Andrews, B. D., Gares, P. A. and Colby, J. D., 2002. Techniques for GIS modelling of coastal dunes. *Geomorphology*, **48**, 289–308.

Anthony, E. J. and Blivi, A. B., 1999. Morphosedimentary evolution of a delta-sourced, drift-aligned sand barrier-lagoon complex, western Bight of Benin. *Marine Geology*, **158**, 161–176.

Antonov, J.I., Levitus, S. and Boyer, T.P., 2005: Steric variability of the world ocean, 1955–2003. *Geophysical Research Letters*, **32**, L12602, doi:10.1029/2005GL023112.

Arnaud-Fassetta, G., 2003. River channel changes in the Rhone Delta (France) since the end of the Little Ice Age: geomorphological adjustment to hydroclimatic change and natural resource management. *Catena*, **51**, 141–172.

Aslan, A. and Autin, W. J., 1999. Evolution of the Holocene Mississippi River floodplain, Ferriday, Louisiana: insights on the origin of fine-grained floodplains. *Journal of Sedimentary Research*, **B69**, 800–815.

Bandyopadhyay, S., 1997. Natural environmental hazards and their management: a case study of Sagar Island, India. *Singapore Journal of Tropical Geography*, **18**, 20–45.

Barile, P. J., 2004. Evidence of anthropogenic nitrogen enrichment of the littoral waters of east central Florida. *Journal of Coastal Research*, **20**, 1237–1245.

Bascom, W. H., 1951. The relationship between sand size and beach slope. *Transactions of the American Geophysical Union*, **32**, 866–874.

Bayliss-Smith, T. P., Healey, R., Lailey, R., Spencer, T. and Stoddart, D. R., 1979. Tidal flows in salt marsh creeks. *Estuarine and Coastal Marine Science*, **9**, 235–255.

Benavente, J., Gracia, F. J., Anfuso, G. and Lopez-Aguayo, F., 2005. Temporal assessment of sediment transport from beach nourishments by using foraminifera as natural tracers. *Coastal Engineering*, **52**, 205–219.

Berendsen, H. J. A., Cohen, K. A. and Stouthamer, E., 2007. The use of GIS in reconstructing the Holocene palaeogeography of the Rhine-Meuse delta, The Netherlands. *International Journal of Geographical Information Science*, **21**, 589–602.

Berkeley, A., Perry, C. T., Smithers, S. G., Horton, B. P. and Taylor, K. G., 2007. A review of the ecological and taphonomic controls on foraminiferal assemblage development in intertidal environments. *Earth Science Reviews*, **83**, 205–230.

Bijlsma, L., 1996. Coastal zones and small islands. *In:* R. T. Watson, M. C. Zinyowera and R. H. Moss (eds) *Climate Change 1995: Impacts, Adaptations and Mitigation of Climate Change: Scientific-Technical Analyses (Contribution of Working Group II to the Second Assessment Report of the Intergovernmental Panel on Climate Change)*. Cambridge University Press, Cambridge, 289–324.

Bindoff, N.L., Willebrand, J., Artale, V., Cazenave, A., Gregory, J., Gulev, S., Hanawa, K., Le Quéré, C., Levitus, S., Nojiri, Y., Shum, C.K., Talley, L.D., and Unnikrishnan, A., 2007. Observations: oceanic climate change and sea level. *In:* S. Solomon, D. Qin, M. Manning, Z. Chen, M. Marquis, K.B. Averyt, M. Tignor and H.L. Miller (eds) *Climate Change 2007: The Physical Science Basis. Contribution of Working Group I to the Fourth Assessment Report of the Intergovernmental Panel on Climate Change*. Cambridge University Press, Cambridge, 385–432.

Bishop, P. and Cowell, P., 1997. Lithological and drainage network determinants of the character of drowned, embayed coastlines. *Journal of Geology*, **105**, 685–699.

Boomer, I., 1998. The relationship between meiofauna (ostracoda, foraminifera) and tidal levels in modern intertidal environments of north Norfolk: a tool for palaeoenvironmental reconstruction. *Bulletin of the Geological Society of Norfolk*, **46**, 17–26.

Boyd, R., Dalrymple, R. W. and Zaitlin, B. A., 1992. Classification of clastic coastal depositional environments. *Sedimentary Geology*, **80**, 139–150.

Bray, M., Hooke, J. and Carter, D., 1997. Planning for sea-level rise on the south coast of England: advising the decision-makers. *Transactions of the Institute of British Geographers*, **22**, 13–30.

Briere, P. R., 2000. Playa, playa lake, sabkha: proposed definitions for old terms. *Journal of Arid Environments*, **45**, 1–7.

Briggs, D., Smithson, P., Addison, K. and Atkinson, K., 1997. *Fundamentals of the Physical Environment* (2nd edn). Routledge, London, 557pp.

Broche, P., Devenon, J. L., Forget, P., de Maistre, J. C., Naudin, J. J. and Cauwet, G., 1998. Experimental study of the Rhone plume. Part I: physics and dynamics. *Oceanologica Acta*, **21**, 725–738.

Brown, B. E., 1997. Coral bleaching: causes and consequences. *Coral Reefs*, **16**, S129–S138.

Brown, B. E. and Ogden, J. C., 1993. Coral bleaching. *Scientific American*, **268**, 44–70.

Bruun, P., 1962. Sea level rise as a cause of shore erosion. *Journal of Waterways and Harbour Division*, **88**, 117–130.

Bryant, E., 2001. *Tsunami: The Underrated Hazard*. Cambridge University Press, Cambridge, 320pp.

Bryant, E., 2008. *Tsunami: The Underrated Hazard* (2nd edn). Praxis Publishing, Chichester, 330pp.

Bryant, E. A. and Haslett, S. K., 2003. Was the AD 1607 coastal flooding event in the Severn Estuary and Bristol Channel (UK) due to a tsunami? *Archaeology in the Severn Estuary*, **13** (for 2002), 163–167.

Bryant, E. A. and Haslett, S. K., 2007. Catastrophic wave erosion, Bristol Channel, UK – impact of tsunami? *Journal of Geology*, **115**, 253–269.

Bryant, E. A. and Price, D. M., 1997. Late Pleistocene marine chronology of the Gippsland Lakes region, Australia. *Physical Geography*, **18**, 318–334.

Bryant, E. A., Young, R. W. and Price, D. M., 1997. Late Pleistocene marine deposition and TL chronology of the New South Wales, Australian coastline. *Zeitschrift für Geomorphologie N. F.*, **41**, 205–227.

Budd, W. F. and Smith, I. N., 1985. The state of balance of the Antarctic ice sheet – an updated assessment 1984. *In: Glaciers, Ice Sheets and Sea Level: Effects of a CO2–induced Climatic Change*. National Academy Press, Washington, 172–177.

Burton, M. L. and Hicks, M. J., 2005. *Hurricane Katrina: Preliminary Estimates of Commercial and Public Sector Damages*. Centre for Business and Economic Research, Marshall University, 12pp. Available at http://www.marshall.edu/cber/research/katrina/Katrina-Estimates.pdf (accessed November 2007).

Carriquiry, J. D. and Sanchez, A., 1999. Sedimentation in the Colorado River Delta and upper Gulf of California after nearly a century of discharge loss. *Marine Geology*, **158**, 125–145.

Carter, D., Taussik, J., Bray, M. and Hooke J., 2000. Regional coastal groups in England and Wales: the way ahead. *Periodicum Biologorum*, **102** (Supp. 1), 215–220.

Carter, R. W. G., 1988. *Coastal Environments: An Introduction to the Physical, Ecological and Cultural Systems of Coastlines*. Academic Press, London, 617pp.

Carter, R. W. G. and Orford, J. D., 1993. The morphodynamics of coarse clastic beaches and barriers: a short- and long-term perspective. *Journal of Coastal Research* (Special Issue) **15**, 158–179.

Carter, R. W. G. and Stone, G. W., 1989. Mechanisms associated with the erosion of sand dune cliffs, Magilligan, Northern Ireland. *Earth Surface Processes and Landforms*, **14**, 1–10.

Carter, R. W. G. and Woodroffe, C. D., 1994. Coastal evolution: an introduction. *In:* R. W. G. Carter and C. D. Woodroffe (eds) *Coastal Evolution: Late Quaternary Shoreline Morphodynamics*. Cambridge University Press, Cambridge, 1–31.

Charman, D. J., Roe, H. M. and Gehrels, W. R., 1998. The use of testate amoebae in studies of sea-level change: a case study from the Taf Estuary, south Wales, UK. *The Holocene*, **8**, 209–218.

Chen, X. and Zong, Y., 1999. Major impacts of sea-level rise on agriculture in the Yangtze delta area around Shanghai. *Applied Geography*, **19**, 69–84.

Church, J. A., Gregory, J.M., Huybrechts, P., Kuhn, M., Lambeck, K., Nhuan, M.T.., Qin, D., Woodworth, P.L., Anisimov, A.O., Bryan, F.O., Cazenave, A., Dixon, K.W., Fitzharris, B.B., Flato, G.M., Ganopolski, A., Gornitz, V., Lowe, J.A., Noda, A., Oberhuber, J.M., O'Farrell, S.P., Ohmura, A., Oppenheimer, M., Peltier, W.R., Raper, S.C.B., Ritz, C., Russell, G.L., Schlosser, E., Shum, C.K., Stocker, T.F., Stouffer, R.J., van de Wal, R.S.W., Voss, R., Wiebe, E.C., Wild, M., Wingham, D.J., and Zwally, H.J. 2001. Changes in sea level. *In:* J.T. Houghton, Y. Ding, D.J. Griggs, M. Noguer, P.J. van der Linden, X. Dai, K. Maskell and C.A. Johnson (eds) *Climate Change 2001: The Scientific Basis. Contribution of Working Group I to the Third Assessment Report of the Intergovernmental Panel on Climate Change*. Cambridge University Press, Cambridge, 639–693.

Cicin-Sain, B., 1993. Sustainable development and integrated coastal zone management. *Ocean and Coastal Management*, **21**, 11–44.

Clark, C., 1993. Satellite remote sensing of marine pollution. *International Journal of Remote Sensing*, **14**, 2985–3004.

Clarke, S. J. and Murray-Wallace, C. V., 2006. Mathematical expressions used in amino acid racemisation geochronology – A review. *Quaternary Geochronology*, **1**, 261–278.

Cooper, J. A. G. and Pilkey, O. H., 2004. Sea–level rise and shoreline retreat: time to abandon the Bruun Rule. *Global and Planetary Change*, **43**, 157–171.

Cremona, J., 1988. *A Field Atlas of the Seashore*. Cambridge University Press, Cambridge, 100pp.

Cunningham, D. J. and Wilson, S. P., 2003. Marine debris on beaches of the Greater Sydney Region. *Journal of Coastal Research*, **19**, 421–430.

Dail, M. B., Corbett, D. R. and Walsh, J. P., 2007. Assessing the importance of tropical cyclones on continental margin sedimentation in the Mississippi delta region. *Continental Shelf Research*, **27**, 1857–1874.

Dalrymple, R. W., Zaitlin, B. A. and Boyd, R., 1992. A conceptual model of estuarine sedimentation. *Journal of Sedimentary Petrology*, **62**, 1130–1146.

Davies, K. H., 1983. Amino acid analysis of Pleistocene marine molluscs from the Gower Peninsula. *Nature*, **302**, 137–139.

Davies, J. L., 1964. A morphogenic approach to the worlds' shorelines. *Zeitschrift für Geomorphologie*, **8**, 127–142.

Davis, R. A., Jr., 1997. *The Evolving Coast* (new edn). Scientific American Library, New York, 233pp.

Dawson, A. G., Musson, R. M. W., Foster, I. D. L. and Brunsden, D., 2000. Abnormal historic sea-surface fluctuations, SW England. *Marine Geology*, **170**, 59–68.

Dawson, S. and Smith, D. E., 1997. Holocene relative sea-level changes on the margin of a glacio-isostatically uplifted area: an example from northern Caithness, Scotland. *The Holocene*, **7**, 59–77.

Day, J. W., Boesch, D. F., Clairain, E. J., Kemp, G. P., Laska, S. B., Mitsch, W. J., Orth, K., Mashrqui, H., Reed, D. J., Shabman, L.,Simenstad, C. A., Streever, B. J., Twilley, R. R., Watson, C. C., Wells, J. T. and Whigham, D. F., 2007. Restoration of the Mississippi Delta: lessons from Hurricanes Katrina and Rita. *Science*, **315**, 1679–1684.

Deford, F., 1999. The Florida Keys: Paradise with attitude. *National Geographic*, **196**(6), 32–53.

Dyer, K. R., 1997. *Estuaries: a physical introduction* (2nd edn). Wiley, Chichester, 210pp.

Edwards, R. L., Gallup, C. D. and Cheng, H., 2003. Uranium-series dating of marine and lacustrine carbonates. *Reviews in Mineralogy and Geochemistry*, **52**, 363–405.

El Raey, M., El Din, S. H. S., Khafagy, A. A. and Zed, A. I. A., 1999a. Remote sensing of beach erosion/accretion patterns along Damietta-Port Said shoreline, Egypt. *International Journal of Remote Sensing*, **20**, 1087–1106.

El Raey, M., Frihy, O., Nasr, S. M. and Dewidar, K. H., 1999b. Vulnerability assessment of sea level rise over Port Said Governorate, Egypt. *Environmental Monitoring and Assessment*, **56**, 113–128.

Elliot, M. and McLusky, D. S., 2002. The need for definitions in understanding estuaries. *Estuarine, Coastal and Shelf Science*, **55**, 815–827.

Environment Agency, 1999. Ice victim Beachy slips into the stormy sea. *Environment Action*, **18**, 5.

Feng, J. L. and Zhang, W., 1998. The evolution of the modern Luanhe River Delta, north China. *Geomorphology*, **25**, 269–278.

Fera, P., 1993. Marine eutrophication along the Brittany coasts – origin and evolution. *European Water Pollution Control*, **3**, 26–32.

Finkl, C. W., 2004. Coastal classification: systematic approaches to consider in the development of a comprehensive scheme. *Journal of Coastal Research*, **20**, 166–213.

Finkl, C. W. and Charlier, R. H., 2003. Sustainability of subtropical coastal zones in southeastern Florida: challenges for urbanized coastal environments threatened by development, pollution, water supply, and storm hazards. *Journal of Coastal Research*, **19**, 934–943.

França, C. A. S. and Mesquita, A. R. de, 2007. The December 26th 2004 tusnami recorded along the southeastern coast of Brazil. *Natural Hazards*, **40**, 209–222.

French, P. W., 1997. *Coastal and Estuarine Management*. Routledge, London, 251pp.

French, P. W., 2007. Deltaic and estuarine environments. *In:* C. Perry and K. Taylor (eds) *Environmental Sedimentology*. Blackwell Publishing, Oxford, 223–262.

Galloway, W. E., 1975. Process framework for describing the morphologic and stratigraphic evolution of deltaic depositional systems. *In:* M. L. Broussard (ed.) *Deltas: Models for Exploration*. Houston Geological Society, Houston, 87–98.

Galvin, C. J., 1968. Breaker-type classification on three laboratory beaches. *Journal of Geophysical Research*, **73**, 3651–3659.

Gehrels, W. R., 1994. Determining relative sea-level change from salt-marsh foraminifera and plant zones on the coast of Maine, U. S. A. *Journal of Coastal Research*, **10**, 990–1009.

Gehrels, W. R., 2000. Using foraminiferal transfer functions to produce high-resolution sea-level records from salt-marsh deposits, Maine, USA. *The Holocene*, **10**, 367–376.

Gehrels, W. R., 2002. Intertidal foraminifera as palaeoenvirnomental indicators. *In:* S.K. Haslett (ed.) *Quaternary Environmental Micropalaeontology*. Arnold, London, 91–114.

Gehrels, W. R., Belknap, D. F. and Kelley, J. T., 1996. Integrated high-precision analyses of Holocene relative sea-level changes: lessons from the coast of Maine. *Geological Society of America Bulletin*, **108**, 1073–1088.

Gilbert, G. K., 1885. The topographic features of lake shores. *Annual Reports of the U. S. Geological Survey*, **5**, 75–123.

Giosan, L., Bokuniewicz, H., Panin, N. and Postolache, I., 1999. Longshore sediment transport pattern along the Romanian Danube delta coast. *Journal of Coastal Research*, **15**, 859–871.

Goodbred, S. L. and Kuehl, S. A., 1999. Holocene and modern sediment budgets for the Ganges–Brahmaputra river system: evidence for highstand dispersal to flood-plain, shelf, and deep-sea depocentres. *Geology*, **27**, 559–562.

Gornitz, V., 1995. Sea-level rise: a review of recent past and near-future trends. *Earth Surface Processes and Landforms*, **20**, 7–20.

Gornitz, V., Rosenzweig, C. and Hillel, D., 1997. Effects of anthropogenic intervention in the land hydrologic cycle on global sea level rise. *Global and Planetary Change*, **14**, 147–161.

Great Barrier Reef Marine Park Authority, 2007. *The Great Barrier Reef Marine Park*. Available at: http://www.gbrmpa.gov.au/ (accessed November 2007).

Guilcher, A., Hallégouet, B., Meur, C., Talec, P. and Yoni, C, 1992. Exceptional formation of present-day dunes in the Baie d'Audierne, southwestern Brittany, France. *In:* R. W. G. Carter (ed.) *Coastal Dunes. Proceedings of 3rd European Dune Congress, Galway, 1992*. Balkema, Rotterdam, 15–23.

Hampton, M. A. and Griggs, G. B. (eds), 2004. *Formation, Evolution and Stability of Coastal Cliffs – Status and Trends*. United States Geological Survey, Professional Paper 1693, 123pp.

Haslett, S. K., 2001. The palaeoenvironmental implications of the distribution of intertidal foraminifera in a tropical Australian estuary: a reconnaissance study. *Australian Geographical Studies*, **39**, 67–74.

Haslett, S. K. (ed.), 2002. *Quaternary Environmental Micropalaeontology*. Arnold, London, 340pp.

Haslett S.K. and Bryant, E. A., 2005. The AD 1607 coastal flood in the Bristol Channel and Severn Estuary: historical records from Devon and Cornwall (UK). *Archaeology in the Severn Estuary*, **15** (for 2004), 81–89.

Haslett, S. K. and Bryant, E. A., 2007a. Reconnaissance of historic (post-AD 1000) high-energy deposits along the Atlantic coasts of southwest Britain, Ireland and Brittany, France. *Marine Geology*, **242**, 207–220.

Haslett, S. K. and Bryant, E. A., 2007b. Evidence for historic coastal high-energy (tsunami?) wave impact in North Wales, UK. *Atlantic Geology*, **43**, 137–147.

Haslett, S. K. and Bryant, E. A., 2008. Historic tsunami in Britain since AD 1000: a review. *Natural Hazards and Earth System Sciences*, **8**, 587–601.

Haslett, S. K. and Curr, R. H. F., 1998. Coastal rock platforms and Quaternary sea-levels in the Baie d'Audierne, Brittany, France. *Zeitschrift für Geomorphologie N. F.*, **42**, 507–515.

Haslett, S. K. and Curr, R. H. F., 2001. Stratigraphy and palaeoenvironmental development of Quaternary coarse clastic beach deposits at Plage de Mezpeurleuch, Brittany (France). *Geological Journal*, **36**, 171–182.

Haslett, S. K., Bryant, E. A. and Curr, R. H. F., 2000a. Tracing beach sand provenance and transport using foraminifera: preliminary examples from NW Europe and SE Australia. *In:* I. Foster (ed.) *Tracers in Geomorphology*. Wiley, Chichester.

Haslett, S. K., Davies, P. and Curr, R. H. F., 2000b. Geomorphologic and palaeoenvironmental development of Holocene perched coastal dune systems in Brittany, France. *Geografiska Annaler, Series A*, **82**, 79–88.

Haslett, S. K., Davies, P. and Strawbridge, F., 1998a. Reconstructing Holocene sea-levels in the Severn Estuary and Somerset Levels: the foraminifera connection. *Archaeology in the Severn Estuary*, **8**, 29–40.

Haslett, S. K., Davies, P., Curr, R. H. F., Davies, C. F. C., Kennington, K., King, C. P. and Margetts, A. J., 1998b. Evaluating late-Holocene relative sea-level change in the Somerset Levels, southwest Britain. *The Holocene*, **8**, 197–207.

Haslett, S. K., Davies, P., Davies, C. F. C., Margetts, A. J., Scotney, K. H., Thorpe, D. J. and Williams, H. O., 2001a. The changing estuarine environment in relation to Holocene sea-level and the archaeological implications. *Archaeology in the Severn Estuary*, **11** (for 2000), 35–53.

Haslett, S. K., Strawbridge, F., Martin, N. A. and Davies, C. F. C., 2001b. Vertical saltmarsh accretion and its relationship to sea-level in the Severn Estuary, UK: an investigation using Foraminifera as tidal indicators. *Estuarine, Coastal and Shelf Science*, **52**, 143–153.

Haslett, S. K., Cundy, A. B., Davies, C. F. C., Powell, E. S. and Croudace, I. W., 2003. Salt marsh sedimentation over the past c. 120 years along the west Cotentin coast of Normandy (France): relationship to sea-level rise and sediment supply. *Journal of Coastal Research*, **19**, 609–620.

Healy, T., 1991. Coastal erosion and sea level rise. *Zeitschrift für Geomorphologie N. F., Supplementband* **81**, 15–29.

Hensel, P. F., Day, J. W. and Pont, D., 1999. Wetland accretion and soil elevation change in the Rhone River Delta, France: the importance of riverine flooding. *Journal of Coastal Research*, **15**, 668–681.

Hooke, J. M. and Bray, M. J., 1995. Coastal groups, littoral cells, policies and plans in the UK. *Area*, **27**, 358–368.

Horsburgh, K. and Horritt, M., 2006. The Bristol Channel floods of 1607 – reconstruction and analysis. *Weather*, **61**, 272–277.

Horsfall, D., 1993. Geological controls of coastal morphology. *Geography Review*, **7**(1), 16–22.

Horton, B. P., Edwards, R. J. and Lloyd, J. M., 1999. UK intertidal foraminiferal distributions: implications for sea-level studies. *Marine Micropalaeontology*, **36**, 205–223.

Houghton, J., 1994. *Global Warming: The Complete Briefing*. Lion Books, Oxford.

Houghton, J. T., Meira Filho, L. G., Callander, B. A., Harris, N., Kattenberg, A. and Maskell, K. (eds), 1995. *Climate Change 1995: the Science of Climate Change (contribution of WGI to the second assessment report of the Intergovernmental Panel on Climate Change)*. Cambridge University Press, Cambridge, 572pp.

Hunt, C. O. and Haslett, S. K. (eds), 2006. *Quaternary of Somerset: Field Guide*. Quaternary Research Association, London, 236pp.

Huybrechts, P., 1990. A 3–D model for the Antarctic ice sheet: a sensitivity study on the glacial-interglacial contrast. *Climate Dynamics*, **5**, 79–92.

Inman, D. and Nordstrom, C., 1971. On the tectonic and morphological classification of coasts. *Journal of Geology*, **79**, 1–21.

IPCC, 2007a. Summary for Policymakers. In: Solomon, S., Qin, S., Manning, M., Chen, Z., Marquis, M., Averyt, K.B., Tignor, M. and Miller, H.L. (eds) *Climate Change 2007: The Physical Science Basis. Contribution of Working Group I to the Fourth Assessment Report of the Intergovernmental Panel on Climate Change*. Cambridge University Press, Cambridge, 1–18.

IPCC, 2007b. Summary for Policymakers. *In:* Parry, M.L., Canziani, O.F., Palutikof, J.P., van der Linden, P.J. and Hanson, C.E. (eds) *Climate Change 2007: Impacts, Adaptation and Vulnerability. Contribution of Working Group II to the Fourth Assessment Report of the Intergovernmental Panel on Climate Change*. Cambridge University Press, Cambridge, UK, 7–22.

Jennings, S., 2004. Coastal tourism and shoreline management. *Annals of Tourism Research*, **31**, 899–922.

Judge, E. K. and Overton, M. F., 2001. Remote sensing of barrier island morphology: Evaluation of photogrammetry-derived digital terrain models. *Journal of Coastal Research*, **17**, 207–220.

Kaplan, I. R., 2003. Age dating of environmental organic residues. *Environmental Forensics*, **4**, 95–141.

Kawabe, M., 2004. The evolution of citizen coastal conservation activities in Japan. *Coastal Management*, **32**, 389–404.

Kay, R. and Alder, J., 1999. *Coastal Planning and Management*. E & FN Spon, London.

Kay, R. and Alder, J., 2005. *Coastal planning and management (2nd edition)*. E & FN Spon, London, 380pp.

Knighton, A. D., Woodroffe, C. D. and Mills, K., 1992. The evolution of tidal creek networks, Mary River, Northern Australia. *Earth Surface Processes and Landforms*, **17**, 167–190.

Komar, P. D., 1976. *Beach Processes and Sedimentation*. Prentice-Hall, Englewood Cliffs, 429pp.

Laborel, J. and Laborel-Deguen, F., 1994. Biological indicators of relative sea-level variations and co-seismic displacements in the Mediterranean region. *Journal of Coastal Research*, **10**, 395–415.

Laborel, J., Morhange, C., Lafont, R., Le Campion, J., Laborel-Deguen, F. and Sartoretto, S., 1994. Biological evidence of sea-level rise during the last 4500 years on the rocky coasts of continental southwestern France and Corsica. *Marine Geology*, **120**, 203–223.

Lambeck, K., 1995. Late Devensian and Holocene shorelines of the British Isles and North Sea from models of glacio-hydro-isostatic rebound. *Journal of the Geological Society, London*, **152**, 437–448.

Larcombe, P. and Carter, R. M., 1998. Sequence architecture during the Holocene transgression: an example from the Great Barrier Reef shelf, Australia. *Sedimentary Geology*, **117**, 97–121.

Leonard, L. J., Hyndman, R. D. and Mazzotti, S., 2004. Coseismic subsidence in the 1700 great Cascadia earthquake: coastal estimates versus elastic dislocation models. *Geological Society of America Bulletin*, **116**, 655–670.

Lian, O. B. and Roberts, R. G., 2006. Dating the Quaternary: progress in luminescence dating of sediments. *Quaternary Science Reviews*, **25**, 2449–2468.

Liu, Y.S., Islam, M.A. and Gao, J., 2003. Quantification of shallow water quality parameters by means of remote sensing. *Progress in Physical Geography*, **27**, 24–43.

Long, A. J., Roberts, D. H. and Wright, M. R., 1999. Isolation basin stratigraphy and Holocene relative sea-level change on Arveprinsen Ejland, Disko Bugt, West Greenland. *Journal of Quaternary Science*, **14**, 323–345.

Long, D. and Wilson, C. K., 2007. *A Catalogue of Tsunamis in the UK*. British Geological Survey: Marine, Coastal and Hydrocarbons Programme Commissioned Report CR/07/077.

Lowe, J. J. and Walker, M. J. C., 1997. *Reconstructing Quaternary Environments* (2nd edn). Longman, Harlow, 446pp.

McBride, R. A., Byrnes, M. R., and Hiland, M. W., 1995. Geomorphic response-type model for barrier coastlines: a regional perspective. *Marine Geology*, **126**, 143–159.

McBride, R. A., Penland, S., Jaffe, B., Williams, S. J., Sallenger, A. H. Jr and Westphal, K., 1991. *Shoreline Changes of the Isles Derniers Barrier Island Arc, Louisiana, from 1853 to 1989*. MAP I-2186 (1:75,000), Miscellaneous Investigations Series, U.S. Geological Survey.

McCullagh, P., 1978. *Modern Concepts in Geomorphology*. Oxford University Press, Oxford, 128pp.

McGuire, B., 2005. Swept away. *New Scientist*, 22 October, 38–42.

Mannion, A. M., 1999. *Natural Environmental Change*. Routledge, London, 198pp.

Masselink, G., 1999. Alongshore variation in beach cusp morphology in a coastal embayment. *Earth Surface Processes and Landforms*, **24**, 335–347.

Masselink, G. and Pattiaratchi, C. B., 1998. Morphological evolution of beach cusps and associated swash circulation patterns. *Marine Geology*, **146**, 93–113.

Masselink, G., Russell, P., Coco, G. and Huntley, D., 2004. Test of edge wave forcing during formation of rhythmic beach morphology. *Journal of Geophysical Research*, **109**, C06003, doi:10.1029/2004JC002339.

Massey, A. C., Gehrels, W. R., Charman, D. J. and White, S. V., 2006. An intertidal foraminifera-based transfer function for reconstructing Holocene sea-level change in southwest England. *Journal of Foraminiferal Research*, **36**, 215–232.

Mather, A., 2007. Arid environments. *In:* C. Perry and K. Taylor (eds) *Environmental Sedimentology*. Blackwell Publishing, Oxford, 144–189.

Mathers, S. and Zalasiewicz, J., 1999. Holocene sedimentary architecture of the Red River Delta, Vietnam. *Journal of Coastal Research*, **15**, 314–325.

Meehl, G.A., Stocker, T.F., Collins, W.D., Friedlingstein, P., Gaye, A.T., Gregory, J.M., Kitoh, A., Knutti, R., Murphy, J.M., Noda, A., Raper, S.C.B., Watterson, I.G., Weaver, A.J. and Zhao, Z.-C., 2007. Global Climate Projections. *In:* S. Solomon, D. Qin, M. Manning, Z. Chen, M. Marquis, K.B. Averyt, M. Tignor and H.L. Miller (eds) *Climate Change 2007: The Physical Science Basis.Contribution of Working Group I to the Fourth Assessment Report of the*

Intergovernmental Panel on Climate Change. Cambridge University Press, Cambridge, 747–845.

Meier, M. F., 1984. Contribution of small glaciers to global sea level. *Science*, **226**, 1418–1421.

Michel, D. and Howa, H. L., 1999. Short-term morphodynamic response of a ridge and runnel system on a mesotidal sandy beach. *Journal of Coastal Research*, **15**, 428–437.

Montserrat, S., Vilibi, I. and Rabinovich, A. B., 2006. Meteotsunamis: atmospherically induced destructive ocean waves in the tsunami frequency band. *Natural Hazards and Earth Systems Science*, **6**, 1035–1051.

Morand, P. and Merceron, M., 2005. Macroalgal population and sustainability *Journal of Coastal Research*, **21**, 1009–1020.

Mortimore, R. N. and Duperret, A. (eds), 2004. *Coastal Chalk Cliff Instability.* Geological Society, London, Engineering Geology Special Publication No. 20, 184pp.

Muzikar, P., Elmore, D. and Granger, D. E., 2003. Accelerator mass spectrometry in geologic research. *Geological Society of America Bulletin*, **115**, 643–654.

NERC, 1991. *United Kingdom Digital Marine Atlas.* Natural Environment Research Council, Swindon.

Newman, W. S. and Fairbridge, R. W., 1986. The management of sea level rise. *Nature*, **320**, 319–321.

Nicholls, R.J., Wong, P.P., Burkett, V.R., Codignotto, J.O., Hay, J.E., McLean, R.F., Ragoonaden, S. and Woodroffe, C.D., 2007. Coastal systems and low-lying areas. *In:* M.L. Parry, O.F. Canziani, J.P. Palutikof, P.J. van der Linden and C.E. Hanson (eds) *Climate Change 2007: Impacts, Adaptation and Vulnerability. Contribution of Working Group II to the Fourth Assessment Report of the Intergovernmental Panel on Climate Change.* Cambridge University Press, Cambridge, UK, 315–356.

Nichols, M. M. and Biggs, R. B., 1985. Estuaries. *In:* R. A. Davis (ed.) *Coastal Sedimentary Environments.* Springer-Verlag, New York, 77–186.

Nield, T., 1999. Beachy feet. *Geoscientist*, **9**(3), 12.

Nigam, R., Saraswat, R. and Panchang, R., 2006. Application of foraminifers in ecotoxicology: Retrospect, perspect and prospect. *Environment International*, **32**, 273–283.

Nittrouer, C. A., Kuehl, S. A., De Master, D. J. and Kowsmann, R. O., 1986. The deltaic nature of Amazon shelf sedimentation. *Bulletin of the Geological Society of America*, **97**, 444–458.

O'Reagan, P., 1996. The use of contempory information technologies for coastal research and management – a review. *Journal of Coastal Research*, **12**, 192–204.

Orford, J. D. and Pethick, J., 2006. Challenging assumptions of future coastal habitat development around the UK. *Earth Surface Processes and Landforms*, **31**, 1625–1642.

Orford, J. D., Forbes, D. L. and Jennings, S. C., 2002. Organisational controls, typologies and time scales of paraglacial gravel-dominated coastal systems. *Geomorphology*, **48**, 51–85.

Orford, J. D., Carter, R. W. G., Forbes, D. L. and Taylor, R. B., 1988. Overwash occurrence consequent on morphological change following lagoon inlet closure on a coarse clastic barrier. *Earth Surface Processes and Landforms*, **13**, 27–36.

Orton, G. J. and Reading, H. G., 1993. Variability of deltaic processes in terms of sediment supply, with particular emphasis on grain size. *Sedimentology*, **40**, 475–512.

Park, C., 1997. *The Environment: Principles and Applications.* Routledge, London, 598pp.

Pascual, A. and Rodriguez-Lazaro, J., 2006. Marsh development and sea level changes in the Gernika Estuary (southern Bay of Biscay): foraminifers as tidal indicators. *Scientia Marina*, **70** (Suppl. 1), 101–117.

Pasqualini, V., Pergent-Martini, C., Clabaut, P., Marteel, H. and Pergent, G., 2001. Integration of aerial remote sensing, photogrammetry, and GIS technologies in seagrass mapping. *Photogrammetric Engineering and Remote Sensing*, **67**, 99–105.

Peltier, W. R., 2002. On eustatic sea level history: Last Glacial Maximum to Holocene. *Quaternary Science Reviews*, **21**, 377–396.

Perillo, G. M. E. (ed.), 1995. *Geomorphology and Sedimentology of Estuaries.* Elsevier, Amsterdam, 471pp.

Perry, C., 2007. Tropical coastal environments: coral reefs and mangroves. *In:* C. Perry and K. Taylor (eds) *Environmental Sedimentology.* Blackwell Publishing, Oxford, 302–350.

Pethick, J., 1984. *Introduction to Coastal Geomrophology.* Arnold, London, 260pp.

Pethick, J., 1993. Shoreline adjustments and coastal management – physical and biological processes under accelerated sea-level rise. *Geographical Journal*, **159**, 162–168.

Pethick, J., 2001. Coastal management and sea-level rise. *Catena*, **42**, 307–322.

Pickering, K. T. and Owen, L. A., 1997. *An Introduction to Global Environmental Issues* (2nd edn). Routledge, London, 512pp.

Populus, J., Moreau, F., Coquelet, D. and Xavier, J. P., 1995. An assessment of environmental sensitivity to marine pollutions – solutions with remote sensing and geographical information systems (GIS). *International Journal of Remote Sensing*, **16**, 3–15.

Porebski, S. J. and Steel, R. J., 2003. Shelf-margin delats: their stratigraphic significance and relation to deepwater sands. *Earth Science Reviews*, **62**, 283–326.

Postma, G., 1990. Depositional architecture and facies of river and fan deltas: a synthesis. *In:* A. Colella and D. B. Prior (eds) *Coarse-grained Deltas.* International Association of Sedimentologists, Special Publication, No. 10, 13–27.

Postma, H., 1961. Transport and accumulation of suspended matter in the Dutch Wadden Sea. *Netherlands Journal of Sea Research*, **1**, 148–190.

Priskin, J., 2003. Tourist perceptions of degradation caused by coastal nature-based recreation. *Environmental Management*, **32**, 189–204.

Pritchard, D. W., 1955. Estuarine circulation patterns. *Proceedings of the American Society of Civil Engineers*, **81**(717).

Queensland Environmental Protection Agency, 2003. *Green Island Recreation Area and Green Island National Park Management Plans.* Queensland Government, 24pp. Available at: http://www.epa.qld.gov.au/publications?id=1208 (accessed November 2007).

Raymo, M. E. and Ruddiman, W. F., 1992. Tectonic forcing of late Cenozoic climate. *Nature*, **359**, 117–122.

Reading, H. G. and Collinson, J. D., 1996. Clastic coasts. In: H. G. Reading (ed.) *Sedimentary Environments: Processes, Facies and Stratigraphy* (3rd edn). Blackwell Science, Oxford, 154–231.

Reed, D. J., 1995. The response of coastal marshes to sea-level rise: survival or submergence? *Earth Surface Processes and Landforms*, **20**, 39–48.

Reimer, P.J., Baillie, M.G.L., Bard, E., Bayliss, A., Beck, J.W., Bertrand, C.J.H., Blackwell, P.G., Buck, C.E., Burr, G.S., Cutler, K.B., Damon, P.E., Edwards, R.L., Fairbanks, R.G., Friedrich, M., Guilderson, T.P., Hogg, A.G., Hughen, K.A., Kromer, B., McCormac, F.G., Manning, S.W., Ramsey, C.B., Reimer, R.W., Remmele, S., Southon, J.R., Stuiver, M., Talamo, S., Taylor, F.W., van der Plicht, J. and Weyhenmeyer, C.E., 2004. IntCal04 Terrestrial radiocarbon age calibration, 26–0 ka BP. *Radiocarbon*, **46**, 1029–1058

Reinson, G. E., 1992. Transgressive barrier island and estuarine systems. *In:* R. G. Walker and N. P. James (eds) *Facies Models: Response to Sea-level Change.* Geological Association of Canada, Waterloo (Ontario), 179–194.

Saleh, A., Al-Ruwaih, F., Al-Reda, A. and Gunatilaka, A., 1999. A reconnaissance study of a clastic coastal sabka in Northern Kuwait, Arabian Gulf. *Journal of Arid Environments*, **43**, 1–19.

Saye, S. E., van der Wal, D., Pye, K. and Blott, S. J., 2005. Beach-dune morphological relationships and erosion/accretion: An investigation at five sites in England and Wales using LIDAR data. *Geomorphology*, **72**, 128–155.

Scott, D. B. and Medioli, F. S., 1978. Vertical zonations of marsh foraminifera as accurate indicators of former sea-levels. *Nature*, **272**, 528–531.

Scott, D. B. and Medioli, F. S., 1986. Foraminifera as sea-level indicators. *In:* O. van de Plassche (ed.) *Sea-level Research: A Manual for the Collection and Evaluation of Data.* Geo-Books, Norwich, 435–456.

Selby, K. A. and Smith, D. E., 2007. Late Devensian and Holocene relative sea-level changes on the Isle of Skye, Scotland, UK. *Journal of Quaternary Science*, **22**, 119–139.

Sellner, K. G., Doucette, G. J. and Kirkpatrick, G. J., 2003. Harmful algal blooms: causes, impacts and detection. *Journal of Industrial Microbiology & Biotechnology*, **30**, 383–406.

Shennan, I., Innes, J. B., Long, A. J. and Zong, Y., 1994. Late Devensian and Holocene relative sea-level changes at Loch nan Eala, near Arisaig, northwest Scotland. *Journal of Quaternary Science*, **9**, 261–283.

Shennan, I., Bradley, S., Milne, G., Brooks, A., Bassett, S. and Hamilton, S., 2006. Relative sea-level changes, glacial isostatic modelling and ice-sheet reconstructions from the British Isles since the Last Glacial Maximum. *Journal of Quaternary Science*, **21**, 585–599.

Simms, A. R., Lambeck, K., Purcell, A., Anderson, J. B. and Rodriguez, A. B., 2007. Sea-level history of the Gulf of Mexico since the Last Glacial Maximum with implications for the melting history of the Laurentide Ice Sheet. *Quaternary Science Reviews*, **26**, 920–940.

Skipperud, L. and Oughton, D. H., 2004. Use of AMS in the marine environment. *Environment International*, **30**, 815–825.

Smith, D. E. and Dawson, A. G., 1990. Tsunami waves in the North Sea. *New Scientist*, 4 August, 46–49.

Smith, D. E., Cullingford, R. A., Mighall, T. M., Jordan, J. T. and Fretwell, P. T., 2007. Holocene relative sea level changes in a glacio-isostatic area: new data from south-west Scotland, United Kingdom. *Marine Geology*, **242**, 5–26.

Smith, D.E., Shib, S., Cullingford, R. A., Dawson, A. G., Dawson, S., Firth, C. A., Foster, I. D. L., Fretwell, P. T., Haggart, B. A., Holloway, L. K. and Long, D., 2004. The Holocene Storegga Slide tsunami in the United Kingdom. *Quaternary Science Reviews*, **23**, 2291–2321.

Smithson, P., Addison, K.. and Atkinson, K., 2002. *Fundamentals of the Physical Environment* (3rd edn). Routledge, London, 627pp.

Stephenson, W. J., 2000. Shore platforms: a neglected coastal feature? *Progress in Physical Geography*, **24**, 311–327.

Stephenson, W. J. and Brander, R. W., 2003. Coastal geomorphology into the twenty-first century. *Progress in Physical Geography*, **27**, 607–623.

Stone, G. W., Liu, B., Pepper, D. A. and Wang, P., 2004. The importance of extratropical and tropical cyclones on the short-term evolution of barrier islands along the northern Gulf of Mexico, USA. *Marine Geology*, **210**, 63–78.

Stowe, K., 1996. *Exploring Ocean Science* (2nd edn). Wiley, New York, 426pp.

Stringer, W. J., Dean, K. G., Guritz, R. M., Garbeil, H. M., Groves, J. E. and Ahlnaes, K., 1992. Detection of petroleum spilled from the MV Exxon Valdez. *International Journal of Remote Sensing*, **13**, 799–824.

Stuiver, M. and Reimer, P. J., 1993. Extended 14C data base and revised CALIB. 3.0 14C age calibration program. *Radiocarbon*, **35**, 215–230.

Suhayda, J. N. and Pettigrew, N. R., 1977. Observations of wave height and celerity in the surf zone. *Journal of Geophysical Research*, **82**, 1419–1424.

Sumatra International Tsunami Survey Team, 2005. *The 26 December 2004 Indian Ocean tsunami: initial findings from Sumatra*. United States Geological Survey. Available at: http://walrus.wr.usgs.gov/tsunami/sumatra05 (accessed 20 November 2007).

Summerfield, M. A., 1991. *Global Geomorphology*. Longman, Harlow, 537pp.

Sunamura, T., 1992. *Geomorphology of Rocky Coasts*. Wiley, Chichester, 302pp.

Swift, D. J. P., 1975. Barrier-island genesis: evidence from the central Atlantic shelf, eastern U.S.A.. *Sedimentary Geology*, **14**, 1–43.

Thomalla, F. and Schmuck, H., 2004. 'We all knew that a cyclone was coming': disaster preparedness and the cyclone of 1999 in Orissa, India. *Disasters*, **28**, 373–387.

Thomson, R. E. and Tabata, S., 1987. Steric height trends of ocean station PAPA in the northeast Pacific Ocean. *Marine Geodesy*, **11**, 103–113.

Titov, V., Rabinovich, A. B., Mofjeld, H. O., Thomson, R. E. and González, F. I., 2005. The global reach of the 26 December 2004 Sumatra tsunami. *Science*, **309**, 2045–2048.

Tokohu University, 2005. *Comprehensive analysis of the damage and its impact on coastal zones by the 2004 Indian Ocean tsunami disaster*. Tohoku University Report, Japan. Available at: http://www.tsunami.civil.tohoku.ac.jp/sumatra2004/report.html(accessed 20 November 2007).

Törnqvist, T. E., Bick, S. J., van der Borg, K. and de Jong, A. F. M., 2006. How stable is the Mississippi Delta? *Geology*, **34**, 697–700.

Törnqvist, T. E., Gonzalez, J. L., Newsom, L. A., van der Borg, K., de Jong, A. F. M. and Kurnik, C. W., 2004. Deciphering Holocene sea-level history on the US Gulf coast: a high-resolution record from the Mississippi Delta. *Geological Society of America Bulletin*, **116**, 1026–1039.

Trenhaile, A. S., 1999. The width of shore platforms in Britain, Canada, and Japan. *Journal of Coastal Research*, **15**, 355–364.

Turner, R. E., Baustian, J. J., Swenson, E. M. and Spicer, J. S., 2006. Wetland sedimentation from Hurricane Katrina and Rita. *Science*, **314**, 449–452.

Umitsu, M., Buman, M., Kawase, K. and Woodroffe, C. D., 2001. Holocene palaeoecology and formation of the Shoalhaven River deltaic-estuarine plains, southeast Australia. *The Holocene*, **11**, 407–418.

Viles, H. and Spencer, T., 1995. *Coastal Problems: Geomorphology, Ecology and Society at the Coast*. Arnold, London, 350pp.

Warne, A. G., Guevara, E. H. and Aslan, A., 2002. Late Quaternary evolution of the Orinoco Delta, Venezuela. *Journal of Coastal Research*, **18**, 225–252.

Warrick, R. A. and Oerlemans, J., 1990. Sea level rise. *In:* J. T. Houghton, G. J. Jenkins and J. J. Ephraums (eds) *Climate Change, the IPCC Assessment*. Cambridge University Press, Cambridge, 257–281.

Warrick, R. A., Le Provost, C., Meier, M. F., Oerlemans, J. and Woodworth, P. L., 1996. Changes in sea level. *In:* J. T. Houghton, L. G. Meira Filho, B. A. Callander, N. Harris, A. Kattenberg and K. Maskell (eds) *Climate Change 1995: the Science of Climate Change (contribution of WGI to the second assessment report of the Intergovernmental Panel on Climate Change)*. Cambridge University Press, Cambridge, 359–405.

Werner, B. T. and Fink, T. M., 1993. Beach cusps as self-organised patterns. *Science*, **260**, 986–970.

Whelan, F. and Kelletat, D., 2003. Submarine slides on volcanic islands – a source for mega-tsunamis in the Quaternary. *Progress in Physical Geography*, **27**, 198–216.

White, K. and El Asmar, H. M., 1999. Monitoring changing position of coastlines using Thematic Mapper imagery, an example from the Nile Delta. *Geomorphology*, **29**, 93–105.

Wigley, T. M. L. and Raper, S. C. B., 1993. Future changes in global mean temperature and sea level. *In:* R. A. Warrick, E. M. Barrow and T. M. L. Wigley (eds) *Climate and Sea Level: Observations, Projections and Implications*. Cambridge University Press, Cambridge, 111–133.

Williams, A. T., 1992. The quiet conservators: Heritage Coasts of England and Wales. *Ocean and Coastal Management*, **17**, 151–167.

Williams, A. T. and Caldwell, N. E., 1988. Particle size and shape in pebble-beach sedimentation. *Marine Geology*, **82**, 199–215.

Williams, A. T. and Simmons, S. L., 1997. Estuarine litter at the river/beach interface in the Bristol Channel, United Kingdom. *Journal of Coastal Research*, **13**, 1159–1165.

Woodroffe, C. D., 1993. Geomorphological and climatic setting and the development of mangrove forests. *In:* H. Lieth and A. Al Masoom (eds) *Towards the Rational Use of High Salinity Tolerant Plants*. Kluwer Academic Publishers, The Netherlands, 13–20.

Woodroffe, C. D., 1995. Response of tide-dominated mangrove shorelines in Northern Australia to anticipated sea-level rise. *Earth Surface Processes and Landforms*, **20**, 65–85.

Yamano, H, Miyajima, T. and Koike, I., 2000. Importance of foraminifera for the formation and maintenance of a coral sand cay: Green Island, Australia. *Coral Reefs*, **19**, 51–58.

Yang, X. J., 2005. Remote sensing and GIS applications for estuarine ecosystem analysis: an overview. *International Journal of Remote Sensing*, **26**, 5347–5356.

Young, I. R., 1999. Seasonal variability of the global ocean wind and wave climate. *International Journal of Climatology*, **19**, 931–950.

Young, R. W., White, K. L. and Price, D. M., 1996. Fluvial deposition on the Shoalhaven Deltaic Plain, southern New South Wales. *Australian Geographer*, **27**, 215–234.

Zong, Y., 1998. Diatom records and sedimentary responses to sea-level change during the last 8000 years in Roudsea Wood, northwest England. *The Holocene*, **8**, 219–228.

Index

(glossary entries are not included in the index)